On the degradation of wood pellets
during pneumatic conveying:
Experiments and DEM-CFD simulations

Bibliografische Information der Deutschen Nationalbibliothek
Die Deutsche Nationalbibliothek verzeichnet diese Publikation in der Deutschen Nationalbibliografie; detaillierte bibliografische Daten sind im Internet über http://dnb.d-nb.de abrufbar.
1. Aufl. - Göttingen: Cuvillier, 2021
Zugl.: (RUB) Bochum, Univ., Diss., 2021

Nonnenstieg 8, 37075 Göttingen
Telefon: 0551-54724-0
Telefax: 0551-54724-21
www.cuvillier.de

1. Auflage, 2021
Gedruckt auf umweltfreundlichem, säurefreiem Papier aus nachhaltiger Forstwirtschaft.

ISBN 978-3-7369-7371-8
eISBN 978-3-7369-6371-9

On the degradation of wood pellets during pneumatic conveying: Experiments and DEM-CFD simulations

Dissertation

zur

Erlangung des Grades

Doktor-Ingenieur

der

Fakultät für Maschinenbau

der Ruhr-Universität Bochum

von

Julian Jägers

aus Dinslaken

Bochum 2021

Dissertation eingereicht am:	01.12.2020
Tag der mündlichen Prüfung:	02.02.2021
Erster Referent:	Prof. Dr.-Ing. V. Scherer
Zweiter Referent:	Prof. Dr.-Ing. habil. Dr. h.c. S. Heinrich

Für
Lisa

Für
meine Eltern

»Respice finem. – Bedenke das Ende.«

Graf Hans Caspar von Bothmer

»Hat man den Willen zu etwas, gelingt es auch, gleich, was einem im Wege steht.«

Lord Robert Baden-Powell

»Ich sage Dir nicht, dass es leicht wird.
Ich sage Dir, dass es sich lohnen wird.«

Art Williams

Vorwort

Diese Arbeit entstand am Lehrstuhl für Energieanlagen und Energieprozesstechnik der Ruhr-Universität Bochum während meiner Tätigkeit als wissenschaftlicher Mitarbeiter.

Mein ganz besonderer Dank gilt Herrn Prof. Dr.-Ing. Viktor Scherer für sein Vertrauen sowie die wissenschaftliche und persönliche Förderung und Unterstützung während meiner gesamten Tätigkeit am Lehrstuhl. Auch verdanke ich ihm die Möglichkeit als Teilnehmer bzw. Begleiter an zwei energietechnischen Exkursionen teilgenommen und auf mehreren internationalen Konferenzen wertvolle Erfahrungen gesammelt zu haben. Zudem danke ich ihm für die Übernahme des Hauptreferats.

Herrn Prof. Dr.-Ing. Stefan Heinrich danke ich für das entgegengebrachte Interesse an meiner Arbeit und die freundliche Übernahme des Koreferats.

Großer Dank gilt auch Herrn Dr.-Ing. Siegmar Wirtz für viele wissenschaftliche Anregungen und den Ansporn bei jeder theoretischen und messtechnischen Problemstellung ins Detail zu gehen.

Zudem möchte ich den Projektpartnern von der TU Hamburg-Harburg sowie vom Holz-Energie-Zentrum Olsberg GmbH für die unkomplizierte und kollegiale Zusammenarbeit danken. Hier ist insbesondere das hohe Engagement von Herrn Abdullah Sadeq und Herrn Hans Martin Behr hervorzuheben.

Allen Mitarbeiterinnen und Mitarbeitern des Lehrstuhls danke ich für die herzliche Atmosphäre und sehr gute Zusammenarbeit. Für unsere fachlichen und persönlichen Gespräche, die nicht nur meine Arbeit, sondern auch den Arbeitsalltag und mich persönlich bereichert haben, möchte ich besonders meinen ehemaligen Bürokollegen Dr.-Ing. Frank Wissing, Dr.-Ing. Christoph Pieper, Dr.-Ing. Frederic Buß und Phil Spatz danken. Herrn Pieper danke ich zudem für die gewissenhafte Durchsicht des Manuskripts. Ein großer Dank gilt auch Sandra Cremer und Anke Mörking nicht nur für ihre ausgezeichnete Unterstützung bei administrativen Fragen, sondern auch für ein offenes Ohr in allen anderen Belangen. Weiterhin danke ich vor allem Thomas Gieselmann aber auch Stefan Ludolph, die mir bei den Auf- und Umbauten der aufwändigen Versuchsstände stets eine große Hilfe waren. Ausdrücklich sei auch der Werkstatt des Instituts für Energietechnik gedankt, die meine Konstruktionswünsche stets zu meiner vollsten Zufriedenheit und wenn erforderlich sehr kurzfristig umgesetzt hat.

Ganz besonders danke ich meinen Eltern, Großeltern und meinem Bruder, die mir jederzeit und natürlich auch während der Promotion mit ihrer bedingungslosen Unterstützung ein großer Rückhalt und Helfer in allen Lebenslagen waren. Zuletzt und vor allem möchte ich meiner wunderbaren Frau Lisa für ihr Verständnis und den Rückhalt insbesondere während der zeitintensiven Bearbeitung dieser Dissertation danken.

Bochum, Februar 2021 Julian Jägers

Abstract

Wood pellets pass various transport and storage steps before they are thermally converted. Not only for industrial but also for domestic purposes they are often pneumatically conveyed, especially during the final delivery step. There, wood pellets undergo various mechanical loads, leading to abrasion, pellet breakage and an increase of fines. The accrued fines cause operational problems and system failures during storage, charging and discharging of pellets and can lead to higher pollutant emissions during combustion. Until now, only a few studies exist related to the mechanisms of wood pellet breakage, especially during pneumatic transport.

In the present thesis, the dependence of wood pellet degradation and fines formation during pneumatic conveying on operating conditions like air and product mass flow or shape of pipe components is investigated. Both the size reduction of the cylindrical pellets during pneumatic transport caused by mechanical impacts and the prevailing pressure losses are analysed experimentally and numerically.
Single particle impact tests are performed in order to investigate the breakage behaviour of wood pellets including the effect of particle length, impact velocity and collision angle. Empirical correlations for the statistical description of the pellets' breakage behaviour are developed, including two key-functions: the so-called selection function (breakage probability) and the breakage function (fragment size distribution). Based on the correlations derived, a numerical degradation model (including particle breakage and fines formation) is developed and implemented into the in-house Discrete Element Method (DEM) code of the Department of Energy Plant and Technology of the Ruhr-University Bochum.

Experimental and numerical investigations are conducted using coupled DEM-CFD simulations (Computational Fluid Dynamics) to obtain detailed insights into flow conditions, particle motion and the mechanical loads on the pellets during pneumatic conveying. Apart from operating conditions like air and pellet mass flow, the bend radius is varied in four steps, supplemented by a 90°-Elbow as worst-case scenario and a straight pipe section for reference. The investigations focus on the resulting particle size reduction and the occurring pressure losses. Detailed insights into particle degradation are obtained from numerical results by statistical analysis of mechanical loads on the conveyed particles through inter-particle and especially particle-wall collisions. Numerical results show good qualitative agreement with the experimentally determined degradation rates and contents of fines produced. The DEM-CFD approach applied reflects the measured prevailing pressure losses with high accuracy.

The degradation model developed allows detailed investigation into wood pellet breakage and fines formation during pneumatic conveying and simultaneously enables the design and dimensioning of pipe configurations and operating conditions to prevent excessive particle size reduction and pressure losses during pneumatic transport.

Contents

Glossary

Latin Glossary

A	Area	$[\mathrm{m}^2]$
a	Cylinder radius	[m]
a_1, a_2, a_t, b_1, b_2, b_3, b_t, c_1, c_2, c_α, d_1, d_2, d_α, e_1, e_2, e_α, t_{10}	Breakage function model parameter	[var.]
a_D, A_V, b_D, B_V, C_V	Selection function model parameter	[var.]
b	Half of particle length	[m]
b	Mass-related fragment frequency ratio	[−]
B	Under size (breakage function)	[−]
C_D	Drag coefficient	[−]
C_L	Lift coefficient	[−]
c_f	Content of fines	[wt.-%]
d	Diameter	[m]
D_i	Bend dimensioning (radius to diameter ratio)	[−]
D_i	Wideness of logistic distribution function	[−]
DU	Mechanical durability	[%]
e_n	Coefficient of restitution	[−]
$\boldsymbol{F}$	Force	[N]
$\boldsymbol{f}_{int}$	Momentum source term	$[\mathrm{N\,m^{-3}}]$
$\boldsymbol{g}$	Acceleration of gravity	$[\mathrm{m\,s^{-2}}]$
h	Drop height	[m]
i, j, k	Imaginary base	
k	Specific turbulent kinetic energy	$[\mathrm{m^2\,s^{-2}}]$
k	Spring stiffness	$[\mathrm{N\,m^{-1}}]$

l	Length	[m]
m	Mass	[kg]
$\dot{m}$	Mass flow rate	[kg s^{-1}]
$\boldsymbol{M}$	Torque	[N m]
MC	Moisture content	[wt.-%]
n	Quantity	[–]
$\boldsymbol{n}$	Normal vector	[–]
N	Counter variable	[–]
p	Pressure	[Pa, bar]
$\boldsymbol{p}$	Particle major axis	[–]
P	Particle	
P	Breakage probability (selection function)	[–]
Δp_0	Pressure loss over conveying line	[Pa]
Δp_1	Pressure loss over pipe component	[Pa]
$\underline{q}$	Quaternion	[–]
$\underline{q}^{-1}$	Inverse quaternion	[–]
q_0	Probability distribution (quantity-based)	[var.]
q_0	Real part of quaternion	[–]
q_1, q_2, q_3	Imaginary part of quaternion	[–]
Q_3	Cumulative distribution (mass-related)	[–]
q_r	Radial displacement	[m]
r	Radius	[m]
$\boldsymbol{r}$	Lever	[m]
R	Rolling radius	[m]
R^2	Coefficient of determination	[–]
$\boldsymbol{RA}$	Rotational Axis	[m]
slr	Solids loading ratio	[–]
t	Time	[s]

$\boldsymbol{t}$	Tangential vector	[−]
T	Temperature	[K]
T	Time	[s]
t_n	Collision time	[s]
$\boldsymbol{u}$	Fluid velocity vector	$[\mathrm{m\,s^{-1}}]$
v	(Impact) velocity (magnitude)	$[\mathrm{m\,s^{-1}}]$
$\boldsymbol{v}$	Particle velocity vector	$[\mathrm{m\,s^{-1}}]$
V	Volume	$[\mathrm{m^3}]$
$\dot{V}$	Volume flow rate	$[\mathrm{m^3\,h^{-1}}]$
v_0	Median impact velocity at $\varphi = 90°$	$[\mathrm{m\,s^{-1}}]$
v_{50}	Median impact velocity	$[\mathrm{m\,s^{-1}}]$
v_{90}	Impact velocity at $\varphi = 90°$	$[\mathrm{m\,s^{-1}}]$
$WTACI$	Wall-particle time-averaged collision intensity	$[\mathrm{N\,m^2\,s^{-1}}]$
x, y, z	Cartesian coordinates	[m]
$\boldsymbol{x}$	Position vector	[m]
x_{cp}	Distance between particle's centres of pressure and mass	[m]

Greek Glossary

α	Particle orientation	[°]
α	Breakage function model parameter	[−]
α_r	Angle of repose	[°]
β	Friction coefficient	$[\mathrm{kg\,m^{-1}\,s^{-3}}]$
χ	Exponential correction factor	[−]
δ	Virtual overlap	[m]
ε	Porosity	[−]
ε	Turbulent dissipation rate	$[\mathrm{m^2\,s^{-3}}]$
η	Dynamic viscosity	[−]
γ	Damping coefficient	$[\mathrm{kg\,s^{-1}}]$

γ	Particle aspect ratio	[−]
Γ	Transport property	[var.]
λ	Darcy friction factor	[−]
μ_C	Coulomb friction coefficient	[−]
μ_r	Rolling friction coefficient	[m]
ν	Poisson ratio	[−]
ω	Specific turbulence dissipation rate	[$\mathrm{s^{-1}}$]
$\boldsymbol{\omega}$	Angular velocity	[$\mathrm{rad\,s^{-1}}$]
φ	Collision angle	[°]
$\boldsymbol{\varphi}$	Orientation vector	[rad]
φ_0	Selection function model parameter	[°]
ϕ	Sphericity	[−]
Φ	Flow property	[var.]
ϱ	Rotational angle of quaternion	[°]
ρ	Density	[$\mathrm{kg\,m^{-1}}$]
ς	Rotational vector of quaternion	[−]
σ	Standard deviation	[var.]
$\bar{\bar{\tau}}$	Fluid viscous stress tensor	[$\mathrm{N\,m^{-2}}$]
θ	Position angle of spatial cross section in bend region	[°]
Θ	Moment of inertia	[$\mathrm{kg\,m^{-2}}$]
ξ	Pressure loss coefficient	[−]
$\boldsymbol{\xi}$	Spring displacement	[m]

Subscripts

0	Before, initial
1	After
$\parallel$	Parallel
$\perp$	Perpendicular
acc	Acceleration

$aero$	Aerodynamic
air	Air
avg	Average
c, $calm$	Calming
$calc$	Calculated
$cell$	Fluid cell
$char$	Characteristic
$Coul$	Coulomb
cyl	Cylinder/cylindrical
$diss$	Dissipative
dyn	Dynamic
el	Elastic
end	End
exp	Experimental
f	Fines
f	Fluid
$frag$	Fragment
h	(Delivery) hose
i, j, k	Index
$init$	Initial
$inlet$	Inlet
mag	Magnitude
max	Maximum
min	Minimum
p	Particle, Pellet
$phys$	Physical
pp	Particle-particle
pw	Particle-wall
rel	Relative
res	Resulting

SD	Spring-Dashpot
sim	Simulated
sph	Sphere, spherical
$stat$	Static
sup	Superficial
WC	Wall collision
x, y, z	Cartesian direction

Superscripts

$'$	Particle-related
$'$	State after rotation/translation
b	Buoyancy
c	Collision
d	Drag
l	Lift
n	Normal direction
$norm$	Normalised
p	Particle
p	Pressure
pf	Particle-fluid
r	Rolling
sph_{eq}	Equivalent sphere
t	Tangential direction
vm	Virtual mass

Dimensionless Numbers

Re	Reynolds number

Acronyms

AEBIOM	European Biomass Association
BF	Breakage Function
BIMSchV	Federal Immission Control Act
BP	Breakage Probability
BPM	Bonded Particle Model
CFD	Computational Fluid Dynamics
CGRID	Contact Grid
CP	Common Plane
CPU	Central Processing Unit
CV	Control Volume
DACH	German-speaking countries
DEM	Discrete Element Method
DEPV	German Wood Fuel and Pellet Association
DPM	Discrete Phase Model
DV	Dalla Valle
EN	European Standard
EPC	European Pellet Council
EU28	28 member states of the European Union
FEM	Finite Element Method
HS	Hölzer and Sommerfeld
IEA	International Energy Agency
ISO	International Organization for Standardization
LEAT	Department of Energy Plant Technology
LPT	Lagrangian Particle Tracking
PVC	Polyvinyl chloride
RANS	Reynolds-Averaged Navier-Stokes
ROI	Region Of Interest
RUB	Ruhr-University Bochum

SC	Spherocylinder
SCylinder	Spherocylinder
SF	Selection function
SN	Schiller and Naumann
SST	Shear-Stress Transport
TC	Test Cylinder
TFM	Two-Fluid Model
udf	User-defined function
UK	United Kingdom
USA	United States of America

Operators

$\overline{}$	Average
$\lvert\ \rvert$	Magnitude
Δ	Size change/step
∇	Nabla operator
∇	Gradient
$^{\partial}/_{\partial x}$	Partial derivative
$^{d}/_{dt}$	Normal derivative
$\boldsymbol{x}$	Vector

1 Introduction

Apart from solar, wind and hydro power, solid biomass is the most commonly used renewable energy resource [10]. Due to its availability and security of supply regardless of seasonal and weather conditions, wood plays a decisive role in the CO_2-neutral energy supply for heat and power generation. Wood pellets belong to the most established and widespread commercial types of woody biomass fuels [54]. The pellets made of dried, mainly untreated wood allow for the value-creating usage of industrial wood and saw mill waste. Due to their good handling and storage characteristics, high energy density and comparatively homogeneous particle sizes and shapes, wood pellets are highly suitable for industrial heat and power generation as well as for decentralised domestic heat generation by automatically fed small-scale combustion systems. Figure 1.1 provides an overview of worldwide wood pellet production and consumption in 2018 compared to 2017.

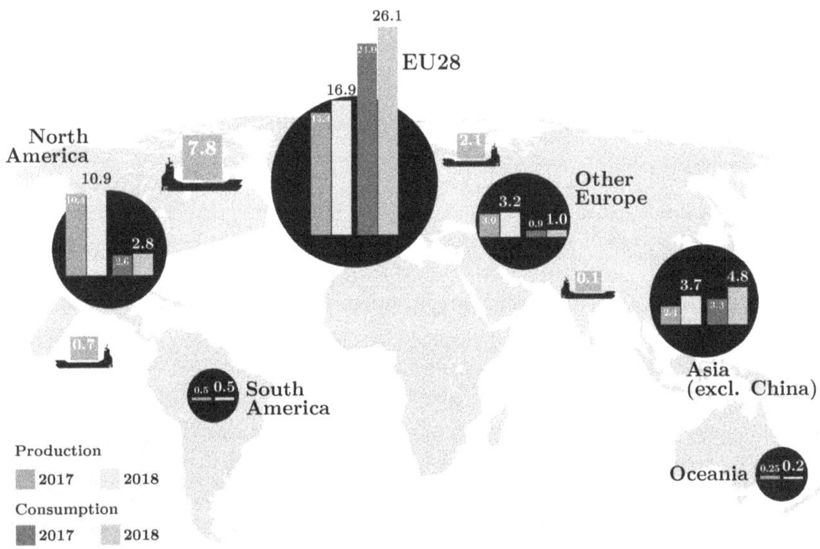

Figure 1.1: World pellet map and trade flow in 2018 (million tonnes) [35]

It confirms the current dominance of the European countries (EU28) concerning wood pellet consumption, but according to statistics of the European Pellet Council (EPC), Asia's market grows rapidly and becomes the driving force behind the development of the global wood pellet market alongside EU28 [35]. The European Union imports wood pellets mainly from USA or Canada and from neighbouring European countries (primarily Russia). The central European demand is significantly driven by industrial consumption in the UK, Denmark and Belgium [174]. Nevertheless, the residential heating sector accounts for 55 % of the required wood pellets, which is the larger share compared to the industrial sector [173]. Within EU28, the EPC expects an increase of woody biomass availability by the growth of fast-growing plants [35]. Note that in the present thesis the word *pellet* is always in reference to *wood pellet* unless otherwise stated.

1.1 Motivation

After pelletising, wood pellets pass various transport and storage procedures. Therein, they undergo different mechanical loads which lead to degradation through abrasion and breakage. Thus, the bulk's content of fines (according to EN 15234-2 particles with a size <3.15 mm [25]) is increased. Especially the transportation steps between the individual storage facilities cause high mechanical loads. According to Mina-Boac et al. [153], wood pellets are transported, retrieved or stored at least eight times on their way from production to combustion. For the route from production in British Columbia to thermal conversion in Europe, Oveisi et al. [167] even assume an average number of ten of so-called handlings. Not only for industrial but also for domestic purposes wood pellets are often conveyed pneumatically. In particular, these processes at various steps of the production and supply chain can cause significant pellet degradation, which is indicated exemplary in experiments by Wiese [233] or Wiese et al. [232], Abdulmumini et al. [3] and in studies of IEA Bioenergy [99].

There exist various reasons for avoiding pellet degradation and thus the increase of fines. In some cases, very small particles are produced and emitted as airborne dust, which can be explosive and is therefore highly relevant for safety aspects. Further, an increased amount of fines favours dust emissions from furnaces, which are potentially harmful to human health and are therefore limited e.g. in Germany by the Federal Immission Control Act (BIMSchV) [159]. The dangers of emission of harmful organic particulate matter and bioaerosols during loading and unloading of pellet delivery vehicles are discussed in the IEA Bioenergy studies [99]. In addition to safety and health aspects, higher amounts of fines play a decisive role concerning undisturbed domestic facility operation. For example, high fines contents can cause blocking of fuel supply devices like screw conveyors or suction probes. Moreover, the accumulation of fines caused by repeated silo loading and unloading processes reduces the domestic storage capacity and thus requires more maintenance [160].

Pneumatic conveying is one of the established transportation methods for wood pellets [54]. A problem in practice is that ideal conveying conditions can rarely be achieved. For example, plug flow conveying, which is considered as the most gentle way of pneumatic transport, is rarely applied due to high risk of blockages and its complex technical implementation during instationary deliveries by silo trucks [151]. Especially the conditions of the final delivery process to domestic storages vary considerably. On the one hand, the conveying conditions depend on local circumstances like hose length, shape and number of bends as well as the storage geometry. On the other hand, the actual operational parameters like air and product mass flows are manually set by operating staff and thus differ as well [48]. Despite the high quality of the mainly EN*plus*-certified pellets (comply high quality standards with a low amount of fines), operational problems can still occur due to generation of fines by inappropriate selection of pipe components or operating parameters during pneumatic conveying.

1.2 Objective and outline

The primary aim of the current thesis is to investigate the dependence of wood pellet degradation and fines formation during pneumatic conveying on operating conditions like air and product mass flow or shape of pipe components. The size reduction of the cylindrical pellets during pneumatic transport caused by mechanical impacts is analysed both experimentally and numerically. For detailed insights into occurring breakage mechanisms, particle-resolved data (e.g. contact frequencies, collision velocities and angles, etc.) are mandatory. As a central numerical tool, the in-house Discrete Element code (DEM) of the Department of Energy Plant Technology (LEAT) of the Ruhr-University Bochum (RUB) is applied. This DEM code determines the particle-particle and particle-wall interactions together with the particles' exchange of momentum with the particle-surrounding fluid phase. For considering the interactions of fluid phase and particles, the DEM code is coupled with a commercial Computational Fluid Dynamics (CFD) solver (ANSYS Fluent®). For numerical assessment of pellet degradation during pneumatic conveying, the existing DEM code is extended by an empirical degradation model, containing statistical functions for breakage probability and size distribution of the resulting fragments. Thus, direct modelling of particle degradation and the subsequent interactions of particles, fragments and the surrounding fluid phase in dependence of varying operating parameters is achieved. The accuracy of the model developed is assessed by comparison of results of coupled DEM-CFD simulations and corresponding experiments performed with varying operating conditions and shapes of common pipe components.
Furthermore, resulting pressure losses, which play a decisive role in design and dimensioning of conveying lines or operating conditions are investigated experimentally and determined numerically, respectively. Thus, important insights are obtained to give recommendations regarding critical components or operating conditions and thus reduce the risk of blockages.
Apart from degradation behaviour and prevailing pressure losses, the numerical approach allows detailed insights into the motion of complex-shaped poly-disperse particles, flow profiles and mechanical wall stresses during pneumatic conveying processes.

The current thesis is divided into seven parts. **Chapter 2** starts with essential basics regarding both wood pellets and pneumatic conveying and gives an overview of the current state of research. **Chapter 3** describes the applied numerical approaches of DEM and CFD. The implemented models of both solid and fluid phase are explained, before the applied momentum-coupling scheme between both phases is discussed in more detail. **Chapter 4** deals with the degradation characteristics of single particles due to collision with a target plate. Single particle impact tests are performed to develop empirical functions for statistical description of breakage probability and size distribution of the resulting fragments in dependence on parameters like initial particle length, collision angle and velocity. The model developed and implemented in the in-house DEM approach serves as the basis for the following numerical simulation of particle degradation effects of pneumatic conveying processes. Experimental and numerical results of the single particle impact tests and thus the validity of the developed breakage model are finally compared or discussed, respectively. **Chapter 5** describes both the experimental and numerical setup for the investigation into pneumatic conveying of wood pellets, focused on the influence of varying conditions on particle size reduction. Furthermore, numerical models and corresponding model parameters applied for determination of inter-particle and particle-fluid interactions are verified. In **chapter 6**, experimental and numerical results of the investigation into pneumatic conveying processes are discussed. The impact of the shape of different pipe components and the influence of varying operating conditions on particle motion and fluid phase behaviour, as well as on the wall or particle stresses (and thus the particle size reduction) are described. Additionally, the suitability of the numerical degradation model developed for predicting size reduction effects during pneumatic conveying is examined on application. In **chapter 7**, the implemented degradation model and the individual numerical and experimental results are summarised conclusively. The thesis finishes with an outlook on further fields of application and potential approaches for improvement.

Within the framework of the current thesis, publications were made [a–j]. The findings of these publications are summarised in this contribution and expanded to unpublished results.

2 Basic principles and state of the art

2.1 Wood pellets as biomass fuel

Requirements for solid biomass for energy generation are basically defined in the standards ISO 16559 [31] and ISO 17225 [26–29, 32–34]. In general, these standards distinguish six classes of biomass derived fuels (wood pellets, wood briquettes, wood chips, logs, non-woody pellets and non-woody briquettes).
According to ISO 16559 [31], wood pellets are defined as a fuel pressed from pulverised biomass with or without additives, usually cylindrically shaped with broken edges and a length between 3.15 and 40 mm. Due to the specifications defined by ISO 17225-2 [33], only pellets with a diameter between 6 and 8 mm are authorised for use in small and medium scaled firing systems. All classes of biogenic solid fuels, which are classified according to ISO 17225-1 [32] on the basis of origin and source, can be used as raw material for wood pellet production. This includes woody or straw-like biomass, biomass of fruits and mixtures of both. In accordance to the focus of the current thesis, figure 2.1 provides only the subclasses of woody biomass.

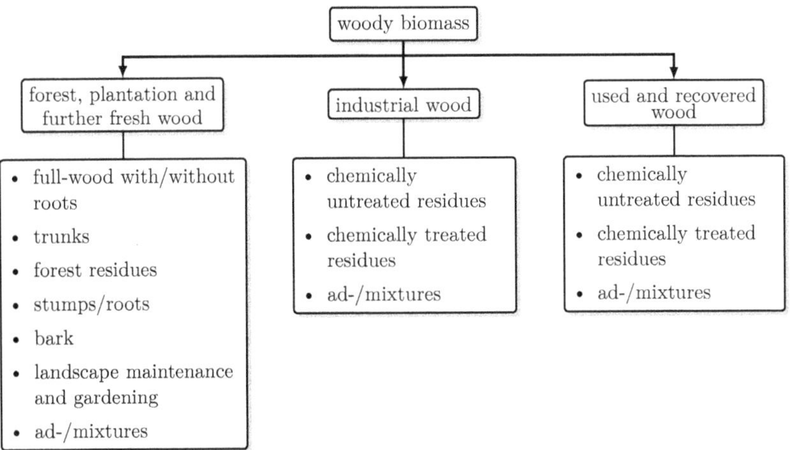

Figure 2.1: Classification of woody biomass according to ISO 16559 [31]

The chemical composition and thus the properties of the pellets vary significantly depending on the pellet mill and its location, the ratio of hardwood to softwood, the mixture of species and chip sizes (branches vs. wood chips vs. sawdust, etc.). For example, a pellet mill in Southampton (Virginia, USA) purchases 100 % of hardwood types, while another mill in Cottondale (Florida, USA) processes mainly softwood types for the production of wood pellets [127]. Pellet plants located geographically close together provide relatively comparable ratios of hard- and softwood [202]. In Central Europe, mainly chemically untreated industrial wood without bark, by-products of the wood-working and processing industry in form of chips, wood flour or cross-cut wood are used for production of wood pellets. Industrial logs are only used as raw material for pellet production in addition to sawmill residues. However, a future increase in demand for biofuels is generally expected (cf. chapter 1), so further biomass potentials have to be developed and the use of round logs will increase [54].

Compared to other biomass fuels, wood pellets can easily be specified and classified. The raw materials available for wood pellet production are regulated by quality standards according to ISO 17225-2 [33]. Based on this standard, a distinction is made between three different qualities EN*plus*-A1, EN*plus*-A2 and EN*plus*-B. These are based on the certification schemes EN*plus* and EN, developed by the German Wood Fuel and Pellet Association (DEPV) and launched in 2010 by the European Pellet Council, which is working under the European Biomass Association (AEBIOM) [69]. The EN*plus*-certification scheme sets certain standards for pellet production, quality assurance, labelling, logistics, intermediate storage and delivery to the customer. EN*plus*-certified pellets must comply with the limits of the properties to be checked. For instance, the certification system sets limiting values for water and ash content, as well as ash softening temperature and the shares of further additives and species:

- **EN*plus*-A1** Pellets from logs or chemically untreated residues with low contents of ash (0.7 %) and nitrogen (0.3 %),
- **EN*plus*-A2** Pellets made of full-wood or forest residues and bark with slightly increased contents of ash (1.2 %) and nitrogen (0.5 %),
- **EN*plus*-B** Industrial wood and chemically untreated used wood with high contents of ash (2.0 %) and nitrogen (1.0 %).

Both ash and nitrogen contents are limited, since the emissions of dust and nitrogen oxides directly correlate with these values. Additionally, an increased amount of ash requires a special ash removal system and generally implies an increased risk of slag formation and corrosion in biomass boilers [37]. However, in all three cases the certified pellets contain a maximum of 1.0 wt.-% of fines when leaving the last step of loading [6]. Due to the limitation of raw materials in the highest quality level (EN*plus*-A1) to debarked round wood and chemically untreated sawmill residues, this class is mainly relevant for domestic purposes. With decreasing quality level, the permitted range of raw materials and the limiting value for the ash content increase, which results in reduced ash softening temperatures. Thus, the use of both quality grades EN*plus*-A2 and EN*plus*-B is mainly limited to the industrial sector [154]. Depending on the season and weather conditions, the water

content of freshly harvested wood can be up to 60.0 wt.-%. Hence, a further important quality criterion for pellets in all three EN*plus*-classes is the prescribed maximum water content of 10.0 wt.-% [33].
Although, from a technical point of view, all biogenic solid fuel classes are suitable for the production of wood pellets, the quality of the pellets is largely dependent on the raw material used. Döring [54] provides a detailed overview of the influences of individual parameters on pellet quality. In summary, the mass fractions of the three biopolymers lignin, cellulose and hemicellulose are decisive for the carbon content of the raw material and thus the energy content related to the dry mass. The content of cellulose and hemicellulose influences the drying properties of the raw material. The proportion of lignin, which is also used as a natural binder, influences the mechanical durability (DU) of the pellets, which is a quality factor within EN*plus*-certification scheme [36].
The DU of wood pellets is defined as the resistance to any size reduction due to mechanical impacts during transport, loading and storage [74]. It is affected by a number of parameters, such as raw material, chip size, production process, water content and the use of binding. The durability is usually determined in a standardised procedure according to EN ISO 17831-1 [30] with a so called tumbling box tester and is defined as the ratio of the sieved mass of the sample before (m_0) and after (m_1) the test procedure:

$$DU = \frac{m_1}{m_0} \cdot 100 \qquad [\%] \qquad (2.1)$$

Commonly, this value deals as a first reference of the amount of fines that may be produced by mechanical impacts during handling [73]. To obtain EN*plus*-A2 and EN*plus*-B-certification, the mechanical durability of the wood pellets must be at least 97.5 %. Even stricter criteria apply for EN*plus*-A1-certification ($DU \geq 98.0\,\%$) [49].

2.1.1 Production process

The pelletising technology used for the production of wood pellets has its origins in Canada and was adapted to European market requirements in the 1990s. As already mentioned, biomass pellets are produced from a wide variety of raw materials and have a wide range of applications today, ranging from individual fireplaces to central heating systems up to the use in industrial power plants [141]. Due to heterogeneous wood species, varying size of the chips and the different water content, the wood used must be prepared before pelletising. The integration of the necessary preparation steps into the process chain of pellet production from sawdust is depicted in figure 2.2 exemplarily.

Depending on the intended application, there exist quality criteria for the final product, which, among others, determine the limits of respective ingredients. In this context, the focus is not only on reducing emissions but, for example, also on preventing corrosion of the firing system. For this reason an initial analysis of the raw material is necessary before production. Furthermore, the production methods used for debarking, drying and crushing depend on the raw material and its composition, e.g. the water content. In Germany, the initial raw materials with an average water content between 35 to 45 % (e.g. sawdust) are mostly dried in belt dryers, which reduces the water content to 10 to 14 % [139]. Subsequently, the dry raw material is crushed within hammer mills to achieve a uniform particle size distribution. For many applications, a comminution next to 1 mm below the end product diameter has been determined as suitable [133].

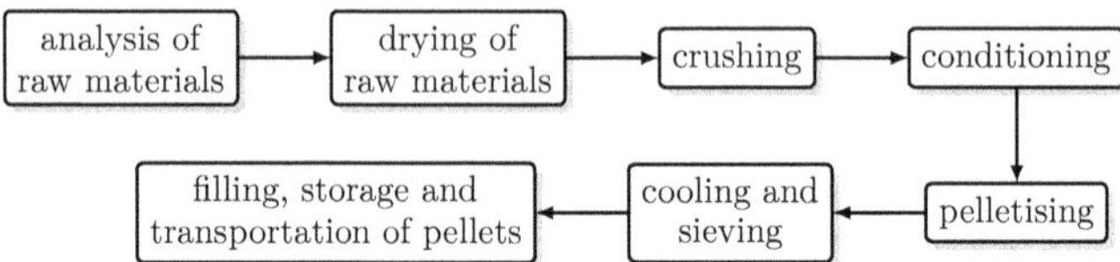

Figure 2.2: Steps of pellet production process [105]

In addition to a uniform product quality, the milling process leads to an increase of the specific raw material surface. Thus, the natural binder lignin can be better split up during pelletising. Next, the biogenic raw material is conditioned to improve the general binding properties. Depending on the water content, the crushed biomass is moistened to ensure a water content between 10 and 15 %. For improving the durability of the final product and reducing friction inside the pelletiser, auxiliary materials or additives can be added to the raw material in the conditioner or even before in the hammer mill. Subsequently, the conditioned raw material is pressed into pellets. For this purpose edge mills with flat or ring dies are usually applied. The pressure inside the edge mill and increased temperatures (up to 298 K) resulting from friction activate the adhesion capability of the lignin, which coats the cellulose fibers and binds the initial raw material to a stable end product. For fast solidification and ensuring dimensional stability, the pellets are cooled down to ambient temperature (about 25°) [140]. Finally, the residual material is separated by vibrating sieves [59].

2.1.2 Domestic delivery and storage procedure

Detailed insights into the entire (global) supply and process chain of wood pellets including raw materials treatment, pelletising process and final delivery steps are provided by Hughes et al. [98], Boukherroub et al. [18] and Uasuf [221]. In accordance to the focal point of the current thesis, the pneumatic delivery and storage procedure in domestic storage facilities is briefly described in the following and illustrated exemplary by figure 2.3. For domestic delivery, wood pellets are usually loaded from storage silos at the plant into blowing trucks and subsequently transported to the customer. With conveying air, provided by the truck's compressor, the pellets are pneumatically blown via hoses through the house ports into the domestic storages. There, they hit an impact protection mat and accumulate at the bottom of the silo. Conveying conditions can widely vary depending on local circumstances or the vehicle-related parameters adjusted by the driver and thus result in different degradation effects, e.g. the hose length, mainly dependent on the distance between truck and house ports. Since the entire hose consists of individual pieces, the number of couplings increases with total hose length, which might have progressive influence on the entire degradation effect [c]. Further mechanical loads occur inside the buildings, e.g. number and geometry of pipe components (bends of different radii, pipe reducers, etc.). In addition to the conveying air, which can be split up inside the truck depending on the vehicle's design (e.g. into the silo cells or for subsequent acceleration) further parameters like pellet mass flow can be set by operating staff. Conclusively, this results in a wide range of possible flow conditions due to varying solids loading ratios. Since the conveying air flow is directly dependent on the hose length (or, in other words, on the pressure loss that needs to be compensated), the vehicle-related operating parameters often depend directly on local conditions.

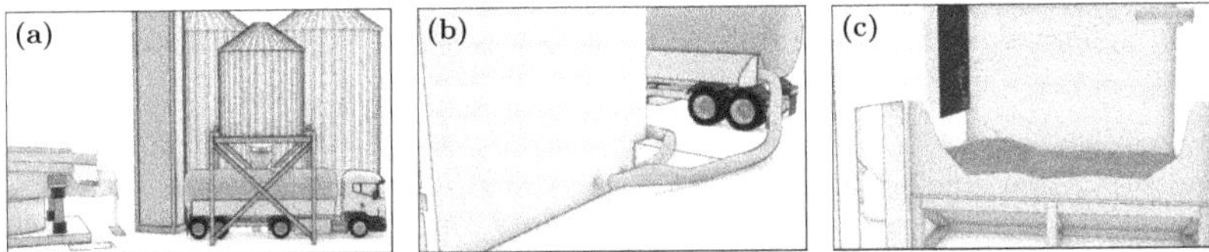

Figure 2.3: Steps of domestic pellet delivery: truck loading (a), pneumatic delivery from silo truck to house ports via hoses (b), silo loading (c) [60]

As an example of the size reduction effects of pneumatic deliveries on the pellets conveyed, figure 2.5 depicts the influence of hose length and conveying air flow on the resulting length distributions. For both hose length (a) and conveying air flow (b) the size reduction effects are clearly indicated by the respective length distribution curves being shifted to the left to shorter pellet lengths and the reduced average particle lengths.

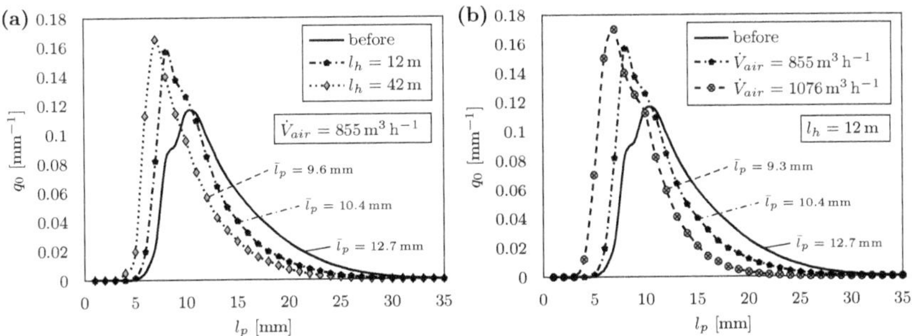

Figure 2.5: Influence of hose length (a) and air volume flow (b) on pellet length distribution during pneumatic delivery with a blowing truck [c]

2.1.3 Pellet degradation and fines formation

General tendencies of particle degradation and thus fines formation due to pneumatic conveying of bulk materials were comprehensively investigated (see section 2.3.4). Nevertheless, the results obtained can't be directly related to the pneumatic conveying of wood pellets [j]. For example, Mina-Boac et al. [153] and Aarseth [2] investigated the pneumatic transport of feed pellets. The authors determined a major effect of the conveying process, e.g. conveying velocities or number of transport repetitions and strong dependence of particle breakage on the durability of the pelletised particles. But the dimensions (3 to 6 mm diameter) and the mechanical durability (DU = 92 to 96 %) are considerably different from those of wood pellets (6 to 8 mm, DU = 98 to 99 %), so that the conclusions of both studies are not completely applicable to the degradation of wood pellets.

The number of both experimental and numerical studies on wood pellet breakage itself is comparatively low. For example, Oveisi et al. [167] investigated the comminution and abrasion effects of wood pellets by free-fall drop tests, taking into account different drop heights and repetitions. As expected, the study revealed that particle degradation increases with ascending drop heights. The durability of the pellets is reduced by repeating drop tests, thus the experienced loads are retained. Free-fall tests, however, represent the situation during pneumatic conveying only rudimentary.
Experiments on pneumatic conveying of wood pellets were performed by Abdulmumini et al. [3]. Their objective was to compare various types of strength testing devices and to determine whether the durability determined by these testers leads to the same trends in fines formation compared to an exemplary pneumatic conveying process. Thus, they were able to determine a general degradation effect of the conveying process, but did not specify the particle comminution or the influence of operating conditions in more detail. Kotzur et al. [114] detected increasing wood pellet size reduction effects with rising pellet length and conveying velocity during lean phase pneumatic conveying. However, theses tests were limited on a sample size of only 100 particles. Again, the impact of pipe components and solids loading ratios on the particles' breakage characteristic was not investigated.
In a recent study, Wiese et al. [232] analysed pellet degradation and fines formation during pneumatic truck deliveries into a model storage. Here, hose length and impact protection mat position were varied. In addition, pellets of different origin were considered. Both longer hose lengths and shorter distances between impact protection mat and blowing connector increase the resulting fines content and thus confirm their influence on pellet size reduction during pneumatic deliveries. The pellet type of lower quality (and durability) leads to higher formation of fines.
In their tests with delivery trucks, Jägers et al. [c] confirmed the progressive influence of hose length and impact mat position on pellet breakage and fines formation. Further, they varied the pressure inside the truck's silo cells as well as the distribution scheme of the conveying air within the vehicle (ratio of driving air flow to acceleration air flow).

Additionally, the product mass flow was varied. The authors came to the result that especially an increased conveying air flow strongly contributes to the size reduction of the pellets conveyed. However, in the scope of this study neither the influence of the shape of pipe components was investigated nor statements about the dependence on the conveying air flow were made.

2.2 Basics of pneumatic conveying

Pneumatic conveying in dilute phase mode can be found in literature since 1924, starting with studies of Gasterstadt [67]. The technique had probably been applied industrially before, as the assembly of the components required is quite simple. With literature going back almost 100 years, the scientific field of pneumatic conveying in dilute phase is extensive and wide-ranging, with many elementary and practical studies, both experimentally and numerically. Klinzing [113] provided a comprehensive historical overview of this technology showing its progress in Europe, Japan and the USA.
Pneumatic conveying is a transportation method for granular bulk solids in pipelines using a gas stream. Pneumatic conveying systems are commonly applied in the chemical, pharmaceutical, agricultural and food industries, in construction and mining, as well as in the energy conversion sector for transportation of pulverised fuels and ashes resulting from incineration in power plants. Bulk solids pneumatically conveyed come in different size classes, starting with particle sizes of powder within the µm-range, up to pellets or chunky material in cm-range. Advantages of this method are the high routing flexibility in horizontal and vertical direction, as well as the compact, weather-independent and dust-tight design. It is also characterised by low investment and maintenance costs, combined with little maintenance requirements. In addition, flow-related conveying allows a high degree of automation and enables simultaneously occurring chemical and physical processes such as drying, mixing and segregation or catalytic processes. Disadvantages are equipment and bulk solids wear and a large energy demand [86]. In practice, it has shown that the length of pneumatic transport is limited to lengths of approximately 3 km [229]. Mills [151] even gives an usual maximum of 1.5 km. This can be attributed to the fact, that the conveying gas expands over the pipeline length due to resulting pressure drops, which leads to high conveying velocities and thus to pressure losses and additional wear. According to Klinzing et al. [111], the conveying distances of average facilities are usually less than 1000 m with conveying capacities between 1 and 400 $\mathrm{t\,h^{-1}}$.

2.2.1 Types of pneumatic conveying systems

A wide range of pneumatic conveying systems are available to cover a large variety of purposes. Usually, most of the systems are conventional, continuously operating, open systems at stationary position. However, innovative batch-wise working and closed systems as well as mobile systems are often used, depending on the material to be conveyed.

In addition, systems can be operated either in positive or negative pressure mode or a combination of both. Combined systems are effectively achieved by staging (arranging two or more compressors in series, in order to achieve higher delivery pressures), which is another option in itself. Figure 2.6 provides an overview of conventional conveying systems with a single air source for any given duty.

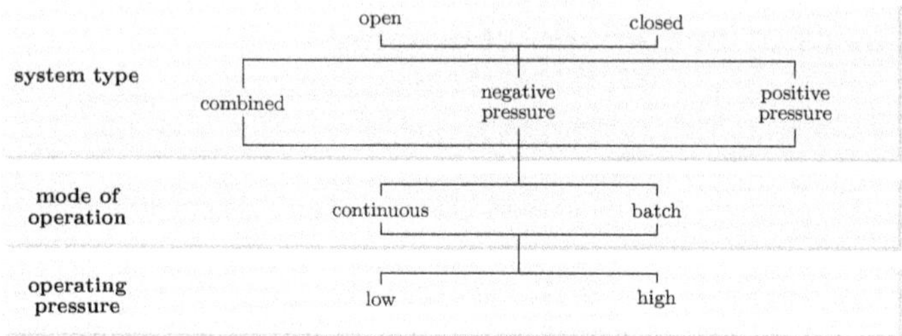

Figure 2.6: Overview of the wide range of conveying systems available for conventional pneumatic transport operating with a single air source

Generally, all provided systems are suitable for conveying of dry bulk particulate materials. For better explanation of the operating modes and fields of application, the systems are subsequently listed in pairs.

- **Open and closed systems** An open system is generally applied where a strict environmental control is not required. Most pipeline systems for pneumatic conveying can achieve completely closed material handling, meaning that with appropriate gas-solid separation and venting, most materials can be handled quite safely in open systems. Many flammable materials are conveyed in open systems by installing necessary safety devices. Commonly, air is used but in some cases nitrogen or other gases are applied. Due to additional operating costs closed systems are commonly used in these specific cases [100].

- **Positive- and negative-pressure systems** Bulk solids can be both sucked and blown, so for pneumatic conveying either pressure or vacuum can be applied. The two types differ fundamentally on how material is fed. By use of diverter valves, delivery to numerous receiving points can be achieved easily in positive pressure systems. Negative pressure systems are often applied for material transportation from multiple sources to a single point [151].

- **Continuously and batch operating systems** Both types are widely used for industrial purposes, e.g. supply of coal dust in combustion applications (continuous) or wood pellet delivery (batch).

- **High- and low-pressure systems** In context of pneumatic conveying, high pressure normally stands for any pressure above 1 bar gauge. In case of systems delivering materials at atmospheric pressure to receiving points, an overhead pressure of 6 bar is usually the maximum due to air expansion problems. Much higher pressures (approx. 20 to 40 bar) are applied to deliver materials to receiving points under pressure in itself, e.g. chemical reactors or fluidised bed combustion systems.

Most pneumatic conveying systems are commonly installed at fixed locations. However, for specific tasks, a variety of mobile systems is available, e.g. for material transport by road, rail or sea. Figure 2.7 depicts the working principle of a mobile system typically applied for road transportation, such as pneumatic delivery of wood pellets by blowing trucks in domestic storages.

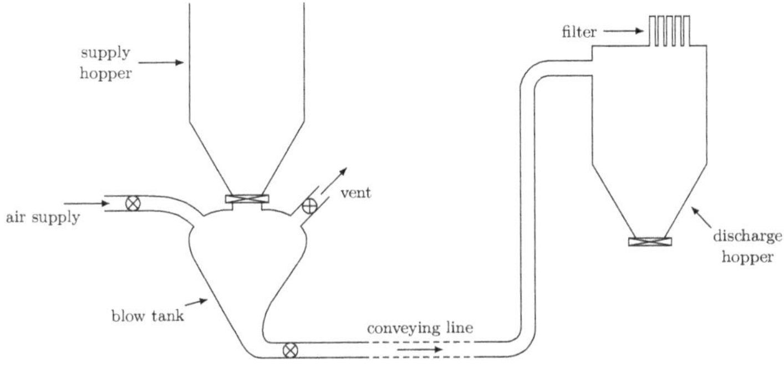

Figure 2.7: Single plug conveying system [151]

Road vehicles are extensively used for transporting a wide range of bulk particulate materials, e.g.: cement, nylon, polyvinyl chloride (PVC) and polyethylene in the chemical industry; sand and soda in the glass industry; and sugar, flour and milk powder in the food industry. Trucks usually have their own on-board positive displacement compressor and thus can unload their bulk solids independently of any depot facilities. The silo on board serves as a hopper for storing the material. It can be lifted to simplify unloading, which can be done via a rotary valve, or the container can be set under pressure for discharging as a blow tank. Both but especially the last option are commonly applied for the delivery of wood pellets [54].

2.2.2 Modes of flow through pipelines

Bulk particulate materials can be pneumatically conveyed in batches or continuously. For large batches or in continuous conveying mode, two types of conveying are indicated, generally defined as *dilute* and *dense phase* flow.

Solids loading ratio For description of the concentration of the conveyed material in the air stream, the dimensionless factor slr (solids loading ratio) is commonly used. It is defined as the ratio of product mass flow $\dot{m}_p$ to air mass flow $\dot{m}_{air}$ used for conveying [155]

$$slr = \frac{\dot{m}_p}{\dot{m}_{air}}. \qquad [-] \quad (2.2)$$

Depending on physical density and particle size of the material conveyed, this parameter can be used to characterise the modes of flow. The state of dilute phase flow results at low slr, whereas in case of high slr, different types of dense phase flows can be assumed. A key benefit of this parameter is that its value remains basically unchanged over the length of a pipeline, contrary to conveying air velocity and volume flow, which continuously change.

Dilute phase flow Conveying materials through pipelines in suspension with the conveying gas is called *dilute phase* conveying or *suspension* flow. Under the assumption that a material can be safely fed into a conveying line, almost any bulk particulate material can be conveyed under dilute phase conditions, regardless of particle size, shape or density. Figure 2.8 depicts characteristic particle flow pattern in a horizontal pipe section.

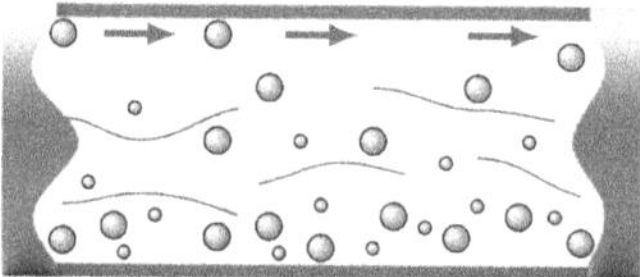

Figure 2.8: Principle of particle flow in dilute phase conveying [85]

According to Mills [151] and Klinzing [112], values of slr up to 15 at typical minimum air velocities from about 11 to 16 $\mathrm{m\,s^{-1}}$ (for fine powdery and granular materials) correspond to dilute phase flow. However, if the pipe section is short and/or the material is conveyed at high pressure, the solids loading ratio can be raised up to 30. Both authors give a variety of examples of slr-ranges for typically pneumatically conveyed bulk materials.
Since dilute phase systems are comparatively straightforward in their design, the initial investment costs are reasonable in comparison to dense phase systems. Further, they are easy and cost-effective to maintain. Due to high conveying velocities achieved in this mode of flow, firstly, significant particle degradation can occur, which results in high formation of dust and fines. Secondly, conveying of abrasive products leads to wear of pipe elements, especially elbows and bends.

Dense phase flow For bulk solids with respective properties, dense phase conveying is possible in two different modes. One of these is generally known as *sliding* or *moving bed* flow and is achievable for materials with good air-retaining properties. The second one, called *plug flow*, is only applicable to materials with good permeability. Here, the conveying air pressure plays a minor role as raised pressure just provides the energy required for conveying the material in dense phase mode over longer pipe sections. The key characteristic of dense phase flows is that the conveying velocity is much lower, i.e. potential wear due to abrasion is much less than for dilute phase flows. The risk of particle degradation and thus losses of product quality is reduced similarly at low conveying velocities [151]. The typical particle flow behaviour in a horizontal pipe section for both modes of flow are depicted in figure 2.9.

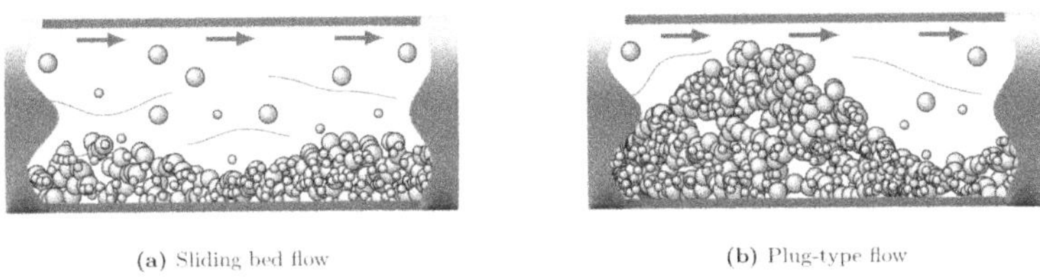

(a) Sliding bed flow (b) Plug-type flow

Figure 2.9: Principle of particle motion in dense phase mode [85]

Sliding bed flow (a) is achievable for bulk solids with effective air retention capabilities, i.e. fine powders of average particle size less than 50 µm (cement, flour, fly ash, etc.) [107]. Most of the material is conveyed on the lower half of the pipe, little dust on top. Vertically, plugs are typically formed and pulsations are probable, but are generally stable. According to McGlinchey [144], solids loading ratios above 100 are common, depending on the product conveyed and air velocities of about $3\,\mathrm{m\,s^{-1}}$ can be achieved, which depicts the economic potential of this flow type.

For plug-type flow (b) the bulk material being conveyed needs high permeability, such as seeds, grains or pelletised products like polyehtylene, nylon or wood pellets. This type of bulk particulate material naturally forms plugs at low conveying air velocities. Plug flows can hardly be characterised by the solids loading ratio in the same way as sliding bed flows mentioned before. Due to the bulks permeability, air easily passes through the plugs so that maximum values of slr in order of 30 can be achieved, even with high pressure drops within the conveying line [164]. In this case, only the conveying air inlet velocity provides a suitable characterisation of the type of flow. Although the solids loading ratio is low, such kind of bulk solids can be reliably conveyed in plug flow at velocities of $3\,\mathrm{m\,s^{-1}}$ [236].

Dense phase conveying has many advantages in comparison to dilute phase conveying. Basically, erosive wear of components and particle abrasion or degradation and thus fines and dust formation are much lower due to lower conveying velocities. Additionally, the specific power requirements for bulk material conveying are comparatively lower [200]. In contrast, dense phase conveying is therefore only possible if the pressure of air supply is high enough or on the opposite, the conveying distance is comparatively short.

Geldart's classification Depending on the material properties, significant overlaps between sliding bed and plug-type flow with regard to conveying velocity and solids loading

ratio can occur. Within each flow regime there exist many variations in flow properties which make the behavioural prediction challenging. However, only few materials are capable of being conveyed in both dense phase flow regimes [151]. The importance of material properties for pneumatic conveying performance has been qualitatively assessed in many studies, e.g. [4, 101, 189, 240]. The most common correlation categorised in terms of particle density and mean particle size is that according to Geldart [68].
The Geldart classification shown in figure 2.10 provides only limited guidance, but was originally derived specifically for fluidisation behaviour in fluidised beds without any reference to pneumatic conveying. Essentially, the classification refers to two material properties. The first one is the difference in densities between the particles' and the surrounding medium, since it is also applicable to bulk solids in liquids. However, for air, the difference can simply be considered as particle density. The second parameter is the average particle size. The classification defines four major areas to identify the behaviour of bulk materials during aeration or fluidisation.

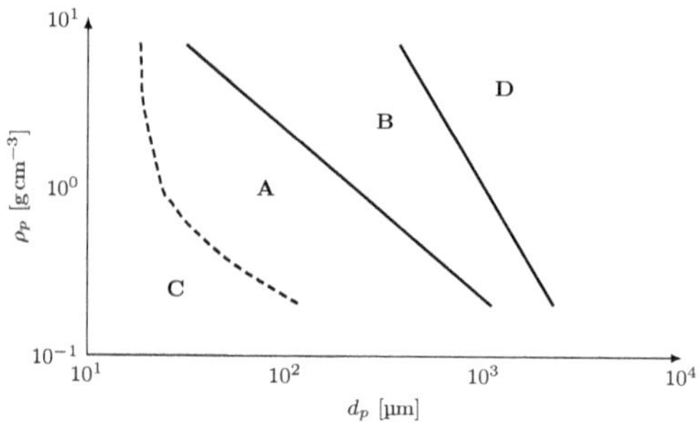

Figure 2.10: Geldart's classification of fluidisation behaviour [68]

The classification is based on the behaviour of a vertical material column when fluidised through a porous base. According to Mills [151], group A materials maintain aeration and the fluidised bed collapses very slowly when air supply has stopped. These materials are most suitable for dense phase conveying. Group B materials do not hold the aeration and the fluidised bed almost collapses after the air supply is switched off. The separation of materials A (air-retaining) and B is about to determine the dense phase conveying capability in the moving bed mode. However, the important property that leads to miss this purpose is the particle size distribution. This makes the division from A to B unreliable and is the reason why the plug flowability cannot be identified for group D materials.
However, due to their material density, particle size and polydisperse size distribution, wood pellets cannot be clearly assigned to one group. Recent fluidisation studies by Agu et al. [4] on wood pellets and wood chips have shown that no reliable classification can be obtained on the basis of solids loading ratio.

2.2.3 Conveying characteristics

Mills et al. [152] defined five parameters which characterise the capability of a pneumatic conveying system for transporting bulk particulate materials: pipe diameter, conveying distance, pressure gradients, conveying air velocity and material properties. The impact of many of these on the conveying behaviour and mode of flow is reasonably foreseeable, but that of the material conveyed is not. Thus, for the design of pneumatic conveying facilities, detailed information about the interaction between the geometry of the conveying line and bulk solids, and in particular the conveying characteristics of the bulk particulate material, are of major significance. Especially the knowledge of the minimum required air velocity as well as the prevailing pressure loss, depending on the material to be conveyed, number and shape of pipe components and conveying distance, are decisive for an efficient and satisfactory operation.

Single phase flow Information on pure air flow through pipelines is essential since the pressure drop required for transporting the air over the length of the pipeline without material provides the reference point characterising the pipeline. The pressure drop of an incompressible fluid (mostly air) with the density ρ_f through a straight conveying line (diameter d, length l) is determined using Darcy's equation [46]

$$\Delta p = \lambda_f \cdot \frac{l}{d} \cdot \frac{\rho_f}{2} \cdot u_f^2, \qquad \text{[Pa]} \qquad (2.3)$$

with the fluid friction coefficient λ_f and velocity u_f.
Pressure losses of pipe bends, elbows or other components are generally described by a coefficient of resistance ξ proportional to the dynamic pressure and are calculated similarly as follows:

$$\Delta p = \xi \cdot \frac{\rho_f}{2} \cdot u_f^2 \quad \text{with} \quad \xi = \lambda_f \frac{l}{d} \qquad \text{[Pa]} \qquad (2.4)$$

An overview of ξ-values of different pipe elements or flow situations is given in [224].
Both equations (2.3) and (2.4) indicate quadratic law-relationship of the pressure drop in relation to conveying velocity. Thus, velocity is a very essential parameter with regard to design or investigation of pneumatic conveying processes. However, a significant issue when using velocity is, that it is not an independent parameter since gases are compressible and their densities depend on both temperature and pressure. Further, the density is reduced with falling pressure, so that the velocity of the carrier gas will increase along the length of a pipeline with a fixed diameter. As the mass flow of the conveying gas will remain independently constant, it is used in the current thesis as substitute for velocity. Assuming an ideal gas, the conveying air velocity can be determined based on the mass flow using the ideal gas law [151].

Gas-solid flows When powdery or particulate material is conveyed in a pipeline at uniform velocity in dilute phase mode, the conveying line's pressure drop increases although the mass flow of the air is kept constant. The level of this increase depends on solids concentration, but of course additionally on the particles' size, shape and density. The dependence of conveying line pressure drop on the bulk particulate solids being conveyed over a wide range of air mass flows and thus velocities for a typically conveyed material (potassium chloride) is provided in figure 2.11.

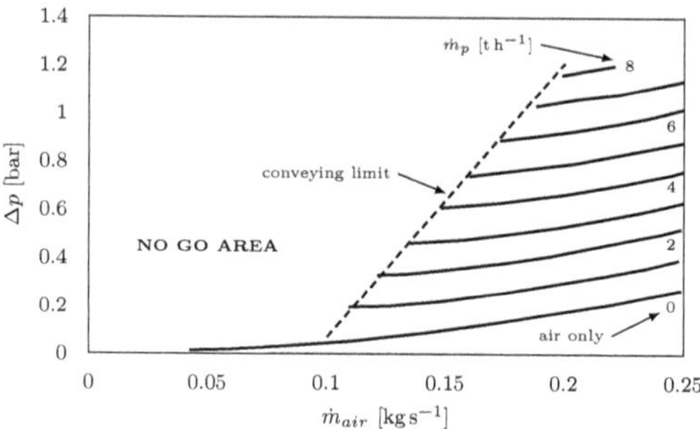

Figure 2.11: Exemplified pressure drop relationship with varying material flow rate [150]

The zero line at the bottom represents the dependence of pressure drop on air flow rate. It therefore provides the lower limit of zero for material conveying capacity for the given system and bulk particulate material. Apart from this, there exist three further limits. Both the maximum limit for conveying air (in combination with material mass flow) and thus the maximum pressure drop limit (on the right-hand side) are actually no limits, since conveying is even possible with higher air flow rates or pressure drops, mainly dependent on the capacities of the installed compressors and material feeding systems. The third and essential limit (on left-hand side) indicates the general safe minimum condition for effective material conveying. There, the lines indeed end and no conveying is possible at lower air mass flows.

However, the correlation between air and product mass flows and the resulting pressure losses are different for each bulk particulate solid and also for each installed conveying system. Thus, the dependencies need to be identified each time individually. For dense phase conveying, there exist different dependencies. Note that these are not covered in the scope of this thesis, but in extensive contributions by e.g. Mills [151], Hilgraf [86] and Klinzing et al. [111].

2.2.4 Bends in pneumatic conveying systems

Bends are mounted in a pneumatic transport facility wherever a directional shift along the conveying line is required. Despite their obvious simplicity (compared to all other parts in pneumatic conveying systems) bends are possibly the least understood and probably most critical components for plant operators [111]. They can roughly be divided into three main categories [50]:

- **Common-radius bends** Long-radius, short-radius, long-sweep bends and elbows.
- **Common fittings** Tees, mitered bends and elbows.
- **Specialised bends** E.g. Vortice Ells, Booth Bends, Diverter Pots, etc.

A broad selection of special bend geometries is provided in figure 2.12. Common-radius bends are manufactured by bending pipes or standard tubes, often with a radius between 1D and 24D (D denotes the pipe diameter) [151]. The conveyed particulate solids may undergo several impacts with the pipe wall or slide along the outer wall of the bend, depending on material properties, solids loading ratio and conveying velocity. At the impact zones, bend wear and product attrition occurs simultaneously.

The most frequently used fitting to achieve changing flow directions is the so-called Blind Tee [50]. Here, one of the outlets is blocked, causing the conveyed solids to accumulate in the pocket (figure 2.12 (c)). The advantage is that the accumulated material damps the impact of the impinging particles, which significantly reduces the resulting wear and product abrasion. The amount of accumulated material in the blind zone varies with bend orientation, solids loading ratio, fluid velocity and material properties (e.g. particle size and density). However, when moving through a tee bend, the conveyed particles lose most of their momentum by impact and therefore must be re-accelerated when leaving the bend. Consequently, the pressure drop caused by a blind tee can be up to three times the pressure drop of a large radius bend [50].

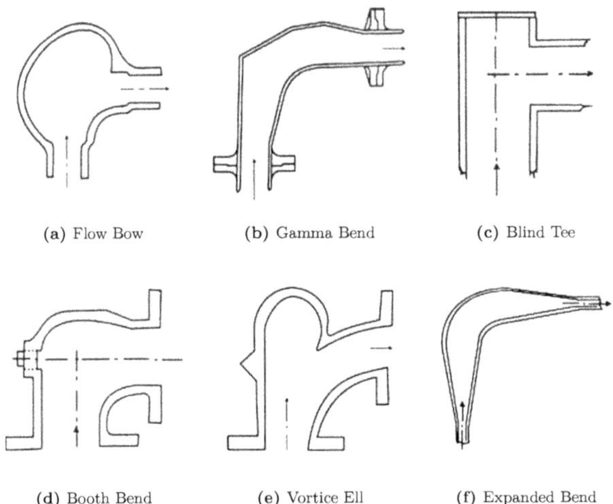

(a) Flow Bow (b) Gamma Bend (c) Blind Tee

(d) Booth Bend (e) Vortice Ell (f) Expanded Bend

Figure 2.12: Special bend designs for pneumatic conveying systems [151]

A variety of special designs are nowadays available to control the flow within the bend to minimise wear and product degradation. This is frequently obtained by creating self-cleaning or refillable pockets or layers of material on which the incoming particles impinge. Wear of the pipe components is reduced by diverting the gas-solid flow away from critical wear points. Exemplary, typical applications include Vortice Ells (figure 2.12 (e)). These have a bulbous extension at the upper end. In dependence of the bend's orientation and the conveying velocity, the material conveyed will either fill the bulbous bowl or circulate in it before exiting. Both cases result in a substantial reduction in wear and product degradation.

The application of specific bends influences the conveying behaviour and thus ultimately the operating conditions required for a gentle pneumatic transport. The major influences are

- pressure drop related to bend,
- attrition or product degradation,
- pipe wear and erosion.

When a gas-solid flow changes direction, e.g. in a bend, this bend certainly serves as a separator of both phases. Due to centrifugal forces, the particles concentrate along the outer wall. The particles may build streaks and undergo several impacts within the bend region depending on material properties, solids loading ratio, conveying velocity and wall interactions. Due to particle-particle and particle-wall interactions and friction along the outer wall, the particles leave the bend region with a lower velocity than at the bend inlet. Energy required for re-acceleration results in an additional pressure drop, which depends on the extent the particles were slowed down during transition. Briefly put, the pressure loss caused by a bend in gas-solid flow is a result of both frictional loss of the bend itself and the momentum required to re-accelerate the particles to steady-state velocity.
Product attrition and degradation during pneumatic conveying is a major concern in industry. Degradation generally refers to formation of "undesirable" fractions of the material being conveyed, which negatively affect its quality [111]. In order to avoid product size reduction during pneumatic conveying, the numerous particle-wall collisions, especially in bends, have to be reduced by (fluidic) design of the bends, both in terms or frequency and, above all, in strength.
However, not only the formation of fines due to particle breakage or chipping but also surface abrasion can lead to operational problems, e.g. dust formation, poor flowability and an increased tendency of caking. In addition, increased surface abrasion and wear of the pipe components mainly decrease the lifetime and are therefore decisive for the maintenance effort of pneumatic conveying lines. Depending on the specific strength of particle and wall materials, either the particles are damaged (attrition or breakage) or the pipe is worn.

2.3 Experimental studies and numerical examinations

Many experimental and numerical research efforts concerning pneumatic conveying of bulk particulate solids in both dilute and dense phase mode are motivated by various key aspects. Due to procedural and design purposes, on the one hand, resulting pressure losses are often the focus of scientific studies. On the other hand, influences of different operating parameters and pipe components, e.g. bends or couplings, on flow profiles and particle motion are also frequently investigated. The effect of properties of the material conveyed, such as particle geometry, plays a significant role as well. Here, the main focus, especially in numerical studies, is the complex modelling of particle-fluid interactions depending on the shape of non-spherical particles. In principle, by investigating pneumatic conveying processes numerically, predictive models concerning fluid or particle moving behaviour and correlations with regard to procedural conditions like pressure drops were developed. For quality assurance and due to operational safety purposes, a further research focus is firstly the wear of pipe components through friction or abrasion and secondly the degradation of the product conveyed.

2.3.1 Experiments on pressure drops and particle velocities

Pressure drop Recent experimental investigations focus on the development of suitable correlations for the description of particle velocities and pressure losses. With respect to the focus of the present thesis, especially the influence of particle properties like geometry of coarse bulk particulate solids is decisive.

The experimental work of Santos et al. [193] partly deals with the particle distribution over the pipe cross-section and partly with the pressure losses occurring during horizontal conveying of polystyrene particles (comparatively large average diameter of 3.2 mm, density of $1050\,\mathrm{kg\,m^{-3}}$). The authors were able to show that despite a suspension flow scheme, polystyrene particles do not present a uniform distribution over the pipe cross-section due to their lump particle size compared to fine particles, but are located in the lower zone of the pipe. This behaviour influences the resulting pressure loss.

Pahk and Klinzing [168] obtained similar results in their comparable investigation of pneumatic dilute phase transport of two different plastic pellet types (cylindrically shaped polystyrene: mean diameter of 3.9 mm, density of $1045\,\mathrm{kg\,m^{-3}}$; elliptically shaped polyolefin: mean diameter of 4.6 mm, density of $900\,\mathrm{kg\,m^{-3}}$). In addition, they were able to demonstrate the dependence of particle motion on its shape. The elliptical polyolefin pellets spin more than the cylindrical polystyrene particles and thus show more irregular motions. During their experiments, the total pressure loss for polyolefin compared to polystyrene particles was higher across all examined solids loading ratios.

Within their contribution, Vásquez et al. [225] were able to provide an explanation to this phenomenon. They analysed the motion of plastic pellets of similar size and density but with different modulus of elasticity during horizontal lean phase conveying visually using high-speed cameras. The authors explained the observed higher pressure drop for softer particles by an increasing number of particle-wall collisions and thus the higher need for multiple re-acceleration of the particles. Due to the resulting lower average particle velocity, the particles' drag coefficient increases, respectively.

Particle velocity Santo et al. [191, 192] investigated the relation between particle and fluid velocity (slip velocity) for pipe cross-sections of different size and a wide range of particle types (various geometry, density, etc.). They developed a predictive model for the description of the dependencies of cross-sectional vertical to horizontal particle velocity using dimensionless correlations. Apart from that, there exist many models for prediction of particle velocities in steady state zones.

However, a decisive aspect in designing pneumatic conveying systems is the understanding of required conditions for stationary states of flow and therefore the necessary length of acceleration pipe sections. Tripathi et al. [214, 216] investigated the acceleration behaviour of various bulk particulate materials (diameter between 0.19 and 4 mm, density between 940 and $5800\,\mathrm{kg\,m^{-3}}$) in suspension flow and developed a predicting model of horizontal acceleration in dependence on different operating conditions. Thereby, they detected that it is generally possible to convey bulk materials at velocities below pickup velocity, respectively.
In recent studies, Zhou et al. [254, 256, 257] examined the acceleration and motion of pneumatically conveyed coarse coal particles (diameter of 5 to 15 mm). The main focus was to develop a model predicting required pickup velocities, depending on various aspects like particle properties or flow conditions. Knowledge of pickup velocities is essential, e.g. to design pneumatic conveying systems without remaining coarse particles at critical line positions. This is mainly avoided by using specialised swirl applications that impose sufficient swirl on the fluid flow.

2.3.2 Numerical investigations

In recent years, in connection with the progress of computer technology, numerical models were widely developed to determine the fundamentals of pneumatic conveying, and increasingly used to evaluate process performance. Approaches considering both solid and gas phase and their interactions can roughly be divided into two categories:

- Continuum-based models for both solid and gas phases (so-called Two-Fluid Model (TFM)), e.g. Euler-Euler.
- Continuum approach for the gas phase and discrete-based consideration of solid particles, e.g. Euler-Lagrange.

Due to its computational simplicity and performance, the TFM concept was applied for investigating pneumatic conveying systems under various conditions, e.g. [14, 79, 84, 132, 134, 143, 169, 175, 176, 264]. The effective application of a continuum model strongly depends on descriptive correlations that are needed to solve the model's governing equations. Currently and despite mixtures of fines and lump particles are often considered, TFM investigations are mostly used for pneumatic transport of powders and fines [14, 79, 84, 143, 169, 175, 176]. According to Kuang et al. [124], it is still challenging to develop TFM-models reproducing all typical flow patterns.

The Euler-Lagrange method is commonly applied as a combined approach using the Discrete Element Method and Computational Fluid Dynamics (DEM-CFD) or the so-called Lagrangian Particle Tracking method (CFD-LPT). The distinction between both is that the last, also known as CFD-DPM (Discrete Phase Model), acts as a kind of reduced DEM-CFD model that doesn't consider particle-particle interactions. Therefore, the CFD-LPT concept is only suitable for pneumatic conveying situations where inter-particle contacts don't play any decisive role [122, 138, 245]. Contrary to the TFM approach, the coupled DEM-CFD model does not require any complex relations between the stress and strain tensors for discrete particles under various fluid flow characteristics and can therefore be applied to a variety of applications. Beyond that, it provides particle-related data like e.g. collision or fluid interaction forces and trajectories, which is essential for understanding the fundamentals behind gas-solid flows [262, 263].

The coupled DEM-CFD approach, for pneumatic conveying was firstly introduced by Tsuji et al. [220] in 1992. It is now widely accepted as an appropriate method for simulating gas-solid flows after being revised by Xu and Yu [97, 241] in 1997 and 1998, respectively. Pneumatic conveying is one of the major areas of application of the coupled DEM-CFD approach. Several phenomena of pneumatic transport have been modelled and analysed, with particular focus on modes of flow, operating conditions, particle degradation, wear of pipe components, heat transfer and electrostatics. A comprehensive overview of exemplifying studies give Kuang et al. [124].

Coupled simulations with non-spherical particles Since non-spherical particles account for 70 % of the raw material in modern bulk handling industries, the particle shape is a decisive factor in the determination of bulk solids behaviour [253]. In recent years, progress towards the development and application of coupled DEM-CFD models for numerical investigation has increasingly been made, especially regarding the influence of particle shape. Modelling of particle-fluid interactions and interaction forces, which is often based on correlation-derived approaches, is a major challenge.

For sake of convenience, the developed numerical models are often compared with experimental fluidised bed models in laboratory scale. For example, pressure losses and fluid velocities, fluidisation schemes, bubble formations, expanding bed volume, detectable void fractions and particle mixing behaviour provide good comparison possibilities between experimental and numerical results. Mahajan et al. [135, 136] varied the size of their model-type fluidised bed in their experimental and numerical studies on the fluidisation behaviour of rod-like particles and observed different fluidisation schemes and corresponding bed heights as well as pressure losses due to the different resulting fluid velocities. In

addition, the authors changed the size of the cylindrical particles and compared their fluidisation behaviour with that of spherical particles.
In their numerical studies, Mema et al. [145] also compared the influence of different modelling approaches describing the drag and lift force acting on non-spherical particles in case of a model-type fluidised bed. They observed differences in the impact of the particle geometry on the resulting particle-fluid interaction forces. Nan et al. [156] compared the influence of different modelling approaches regarding the rod-like particle geometry on resulting fluid velocities, the voidage and the minimum required fluidisation velocity. Further, the authors detected strong dependencies of the geometric particle relations on the resulting probability distribution of the particles' orientation.
Oschmann et al. [165] and Vollmari and Kruggel-Emden [227] investigated the motion behaviour of non-spherical particles numerically, in particular the mixing and the residence times in a model-type or double-chamber fluidised bed and compared the numerical results with experiments. Spheres, cylinders and cuboids of varying proportions, partly modelled as sphere-clusters, were applied. In principle, the authors confirm a good agreement between numerically and experimentally determined mixing rates, particle orientations and residence times. However, results obtained indicated some deviations with respect to the residence time in the double-chamber fluidised bed, since certain particle-wall interactions were not completely modelled.

There exist many efforts to examine numerically the developed and appropriately validated models for investigating the dependence of pneumatic conveying processes on particle geometry. In a numerical study, Oschmann et al. [166] investigated the mixing behaviour of particles of different shape during pneumatic transport. Cylinders as well as plates or cuboids were applied. The results obtained show that mixing or basically particle motion depends on particle-wall interactions as well as on rope formation or dispersion.
On the basis of prolate and oblate ellipsoids or cuboids, Hilton and Cleary [87] analysed the impact of particle geometry on the mode of flow. They observed that spherical or nearly spherical particles tend to convey in a slug flow at high fluid velocities, whereas non-spherical particles remain in dilute phase mode.
A comparison of the flow behaviour between spherical and cylindrical particles was carried out by Ebrahimi et al. [57], adjusting the drag correlations for non-spherical particles. By comparison with experimental results, the authors identified significantly larger deviations for the cylindrical particles, which they, like others, attribute to more irregular particle-wall or inter-particle contacts and the respective underlying drag models.
Significantly larger and more irregular shaped particles (coal particles with sizes between 5 and 30 mm) were involved in a study by Yang et al. [243]. The authors investigated the suspension behaviour of these particles, numerically approximated by sphere-clusters, during vertical pneumatic transport and detected strong dependencies of the minimum suspension velocity on the size of those irregularly shaped coal particles.

2.3.3 Flow through bends

Tripathi et al. recently carried out experimental investigations on the influence of different bend radii and orientations (horizontal-vertical and vice-versa) as well as of type of conveyed bulk particulate materials (e.g. silica gel, fine plastics, sand, pasta flour, etc.) on the particle velocity [217] and the resulting pressure losses [215]. Their primary aim was to determine the minimum required re-acceleration distance behind the bend and to specify the corresponding pressure losses in more detail. Results in [218] for fly ash indicate higher particle velocity loss as well as higher pressure losses (and thus higher energy demand), if the superficial air velocity is higher and bend radii are smaller [104]. Several numerical studies on erosion and wear of pipeline bends due to pneumatic transport (e.g. [204, 242, 256]) exist. A recent theoretical study is performed by Laín and Sommerfeld [130]. The authors investigated wear effects within a bend (horizontal-vertical) for mono- and polydisperse glass spheres depending on conveying velocity and pipe diameter. They identified that more inter-particle collisions reduce erosion, although the frequency of wall collisions is generally increased. In contrast, a particle size distribution with the same number-average diameter as for monodisperse particles leads to a much higher erosion depth. Finally, increasing the solids loading ratio reduced the bend erosion due to changes in particle impact velocity and angle, although the wall collision frequency will increase again.

Chu and Yu [41] carried out coupled DEM-CFD simulations to analyse the mechanical wall loads of pipe bends by pneumatically conveyed particles. Their work also focused on the understanding of rope formation and the dynamics of particle velocities. Their investigations clearly indicated not to neglect the influence of the particles on the gas phase when analysing pneumatic conveying processes.

Kruggel-Emden and Oschmann [117] investigated the influence of particle shape on rope formation as well as the particle motion and orientation during dilute phase conveying by DEM-CFD. By applying cubes, pyramids, plates and icosahedrons, they were able to demonstrate that particle shape has a considerable impact, especially on the particle-particle or particle-wall but also on the particle-fluid interactions.

Schallert and Levy [195] and Du et al. [56] dealt primarily with rope formation caused by multiple or single bends. In both studies, particle behaviour in and behind the bend region was the focal point of the investigations. It was proven that the order (horizontal-vertical or vice-versa) is not insignificant but shows a clear influence on rope dispersion. Pipe erosion effects are the subject of a study by Uzi et al. [222] with spherical particles, in which the authors investigated wear effects depending on the kind of particle-wall contacts. For this purpose, they determined contact statistics (probability density of impact angles, velocities or forces) of the particle-wall collisions in dependence on particle bend inlet velocities. The study resulted in the development of a one-dimensional erosion model, which allows the prediction of erosion with much less numerical effort compared to coupled DEM-CFD approaches.

2.3.4 Particle degradation

Particle breakage occurs - whether intended or unintended - in various processes, e.g. milling [38, 129], fluidised beds [16] or pneumatic conveying [66]. The degradation behaviour is investigated and/or modelled both experimentally and numerically, either on a single particle or process level. The Finite Element Method (FEM) is often applied for detailed numerical investigation or modelling of single particle fracture mechanisms. Here, the particles break into individual fragments depending on their separation by the finite elements. This approach is applied e.g. in [109, 110, 210, 250]. However, in many cases the particles are represented in DEM via clusters of spheres with solid bonds, the so-called bonded particle model (BPM) [7, 8, 55, 71, 206, 255]. The actual particle breakage is modelled by bond removing. In these simulations, the number, size and properties of both bonds and spherical particle elements are used to represent the internal particle structures. Another way in DEM is the representation of particle degradation by splitting an existing particle into individual smaller fragment particles [126, 128, 209]. The individual DEM particle is divided or replaced by fragments at the appropriate positions according to the corresponding breakage criteria.

On single particle level, complex particle shapes are applied for investigating the influence of the particle geometry on comminution behaviour. For modelling particle crushing processes, spherical particles are typically used for simplification to keep the numerical effort comparably low. However, some recent studies also deal with breakage processes of non-spherical, so-called high aspect ratio particles (e.g. needle-shaped or cylindrical particles).
Guo et al. [77, 78] and Grof et al. [75, 76] simulated uniaxial compression and thus the breakage of rod-like particles, using blackboard chalk sticks or smaller scale cylinders (particle length up to 20 mm). The primary concern was to model particle breakage using so-called daughter distribution functions. Results were verified experimentally and fitted with population balance equations. In all cases, the particles were modelled via sphere clusters.
Kumar et al. [126] simulated a similar uniaxial compression scenario, but applied a different approach for modelling the needle-shaped particles and thus for the description of the crushing behaviour. They represented the particles as single spherocylinders (cylindrical shaped particles with hemispherical caps at both ends), which were replaced by fragments after breakage. In their work, the authors focused on the development of functions describing the particles' breakage behaviour. In all cases, there was a good agreement between numerically predicted and experimentally determined degradation characteristics. Within the uniaxial compression scenario, all authors found that breakage effects were progressed by higher lid pressure and increasing inter-particle or particle-wall friction.

For numerical investigation of comminution effects in multiphase systems, coupled DEM-CFD simulations are suitable. Therein, the degradation modelling can be applied in two different ways:

- **Post-processing breakage modelling** Statistics on particle-particle and particle-wall collisions are collected and used to simulate particle degradation applying further sub-models [23, 55, 178].
- **Direct breakage simulation** When a predefined breakage condition is reached, a large particle is split into new smaller particles [103, 128, 209].

The direct breakage simulation approach requires a very high computational effort, since the insertion of fragments leads to an increase of modelled objects. Nevertheless, this method allows more detailed simulations, since the effects of resulting fragments and thus an increased number of particles on further motion behaviour of the bulk particulate solids (and the fluid phase, respectively) can be considered and investigated. In both cases, empirical models based on experimentally determined statistics are typically implemented for breakage description (e.g. [22, 39, 103, 171, 172, 183]).

Due to possible high particle velocities, particle-particle or particle-wall collisions lead to particle breakage, especially in lean phase flow and additionally favoured by bends. The respective influence of impact velocity and angle on the degradation behaviour was therefore specifically investigated using single particle impact tests by Salman et al. [187, 188]. The authors investigated the influence of impact velocity, collision angle, particle size and number of repeating impacts on the resulting percentage of unbroken particles. In addition, simulations combining particle trajectory with particle fragmentation properties were performed to determine the percentage of unbroken particles in a horizontal pipe and bends for dilute phase conveying scenarios. The validation of this predictive model showed a good agreement with the experimental results. The coarse spherical particles used were produced from a general purpose fertilizer (comparatively weak material). Particles with a diameter of 3.2, 5.15 and 7.1 mm with a grain size between 40 and 120 μm were investigated. The authors were able to show that a threshold velocity exists at which no comminution occurs. Regarding the influence of impact angle, it was found that the probability of particle breakage decreases significantly with decreasing angle. Ultimately, this leads to the fact that almost no degradation occurs in straight conveying sections.

The numerical degradation method is based on the model of two types of comminution functions: the process functions (also known as machine functions) and the material functions [65]. Machine functions describe the mechanical loads to which the particles are exposed to during comminution, e.g. impact velocity, collision angles and number of collisions. Material functions define the influence of material properties on the particles' breakage behaviour. This model presumes that on the one hand the machine functions depend on the application and vary for the respective system. On the other hand, the material functions do not depend on the process but only on material properties. Kalman et al. [103] proposed an algorithm in which both the machine and material functions are brought together. The concept is sketched in figure 2.13.
Based on initial size and strength distribution functions (obtained from coupled DEM-CFD simulations), the algorithm is initiated with a random size and strength for each

particle. Afterwards, the machine functions specify the particles' stress state (*stress analysis*). Second, the material functions (*strength properties*) are included in the numerical simulation to describe the individual particle's mechanical stress. An applied force on each particle is compared to its strength, respectively. Particle breakage occurs when the force applied on the particle is higher than its strength (*selection function*). The number and size distribution of the so-called daughter particles is set by a *breakage function*. All daughter particles get a suitable new strength based on the strength distribution function (*initial strength*). A particle will not break if the force applied is less than its strength. However, the strength is reduced due to fatigue following the implemented *fatigue function*. Finally, the process starts again with the next collision event. The authors intended to verify the simulation with a series of *experiments*.

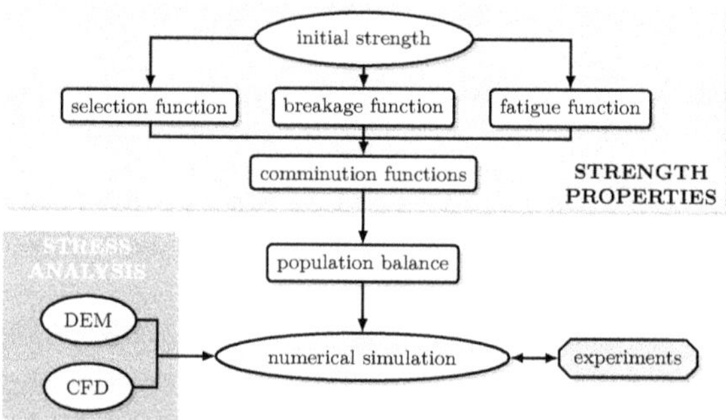

Figure 2.13: Computational breakage algorithm by Kalman et al. [103], combining material and machine functions

Brosh et al. [22, 23] developed a DEM-CFD degradation model based on the work of Kalman et al. [103] described above (cf. figure 2.13). Therein, both the machine functions and the breakage algorithm were calculated by the DEM-CFD approach applied. Regarding the particle's comminution, precise predictions could be made by this model based on the respective empirical distribution functions. Due to the corresponding experimentally determined empirical breakage correlations, the models are only valid for spherical and homogeneous structured particles, which may have a similar breakage behaviour.

This model was extended by Uzi et al. [223], where the stress analysis was performed in advance to define the appropriate machine functions. These include the collision frequency function, which describes the time distribution between the individual collisions, and the collision velocity function, which defines the distribution of collision impact velocities. If these functions are known or developed beforehand (which is actually a must in this case), this model allows for a significant lower numerical effort. The authors observed that smaller particles collide with the pipe wall at higher impact velocities and, in addition, more frequently, while particle-particle interactions occur more frequently for larger particles.

This model was applied and again extended by further machine functions by Portnikov et al. [172] to investigate the comminution effect of bends during pneumatic conveying. Their model predicts the particle degradation effects based on empirical correlations for both machine and material functions and is no longer based on time-consuming DEM-CFD simulations. Based on previous empirical and material dependend studies, the material functions defined include five empirical comminution functions: strength distribution [186], selection (breakage probability) [171, 183], breakage [183], fatigue [185] and equivalence (conversion of the particle's impact load to the force causing the same percentage of broken particles) [184]. The machine functions depend on three empirical and geometry-based functions. The first is the stationary particle velocity distribution function, which was developed empirically by Santo et. al [191] for various materials and particle sizes. The remaining two functions are provided by Portnikov et al. including impact angle distribution and particle position distribution functions. In their work, grains of salt with a particle size between 2 and 2.36 mm were used. Again, the current model is only suitable for homogeneous, uniformly breaking bulk particulate solids.

In the scope of the present thesis, especially the functions concerning the particle's breakage probability (selection function) and the breakage function are important. The latter for describing the resulting size distribution of the daughter particles. Rozenblat et al. [183] developed breakage probability and breakage functions depending on material properties and impact velocity, based on numerous single particle impact tests. They applied salt, potash, graphite nanoplatelets and zeolite and developed or compared several approaches describing the particle degradation behaviour in relation to the varied impact velocity but neglecting the impact angle.
Similar single particle impact test were performed by Portnikov et al. [171] to investigate the influence of impact velocity and collision angle on the breakage behaviour of different particle types. In this study, the authors developed a breakage probability function predicting the particle breakage by impact depending on mother particle size, impact velocity and angle. Materials such as salt, potash, silica gel, zeolite or potassium sulfate (spherical particles with sizes from 0.71 to 4 mm) were applied in these studies. The authors shot 50 g of each material against an impact plate and determined the resulting degradation rate by weight. Experimental data show, as expected by the authors, that smaller particles are stronger due to their fewer impurities and cracks. Furthermore, the straight collision causes the maximum damage due to the maximum change in momentum caused by the impact and consequently the maximum collision force. The data indicated that the damage is approximately the same between 50 and 90°, while for angles below 50° the particle damage decreases significantly.

In contrast to all previous studies dealing with spherical particles, Aarseth [2] investigated the comminution effects of pneumatic conveying processes on cylindrical feed pellets. The author determined the dependence of particle size reduction on air velocity, bend radius and number of repeated impacts using commercially available feed pellets in a 100 mm-bore pipeline. Although the mechanical strength of each type of pellets vary, the particle degradation increases exponentially with the conveying air velocity for all three

pellet types. Additionally, short radius bends cause more product damage than longer radius bends at all air velocities tested.
Of course, there exist further experimental studies concerning product degradation. Since most conveyed particulate solids are roughly spherical (except of pellets), all of these focus on spherical shaped particles (e.g. [47, 50, 65, 115, 228]). Since the results and findings, especially in case of dilute phase conveying, are similar to those of Aarseth, a further detailed description is not given.

2.4 Demarcation of own thesis

The literature review reveals that pneumatic conveying as an important transport method for bulk material in industry has been widely examined experimentally and numerically. However, most of the studies concentrate on fine particles which can be approximated as spheres. For these fine, spherical particles, pressure loss, particle breakage and the interrelation of particle breakage and pipe geometry has been studied. Breakage models have been developed which can be included into numerical simulations. However, these models do not consider how many and which fragments are formed. Pneumatic conveying of feed pellets has also been studied to a certain extent, because of their commercial importance, however their material characteristics are not comparable with wood pellets.

A thorough examination of the breakage behaviour of wood pellets during pneumatic conveying is still missing, a gap which will be closed by the current thesis. In particular, the following topics will be addressed:

1. Single particle impact tests which form the basis of the formulation of breakage models which provide breakage probability and the split between fragment and fines,

2. Experimental pneumatic conveying tests at different operating conditions and flow regimes under the variation of pipe component geometry, which will allow for the determination of pressure loss and breakage characteristics (size distribution of fragments and amount of fines),

3. Numerical simulations (DEM-CFD) of the experiments in topic 2 by using the breakage models derived form single particle impact tests in topic 1,

4. Analysis and comparison of experimental and numerical data to finally devise rules and recommendations for process conditions and pipe components which enable a gentle pneumatic transport of wood pellets with minimum comminution.

3 Numerical methods

In the current work a DEM-CFD approach was applied, whereby the particles are represented discretely and the fluid phase is represented as a cell-averaged continuum [44, 219]. The coupled DEM-CFD approach is based on the in-house DEM code of the Department of Energy Plant Technology of the Ruhr-University Bochum. The single-particle-oriented methodology of DEM allows to describe the particle motion within the numerical domain and their interactions with the surrounding environment three-dimensionally and time-resolved. To model the particle-fluid interaction, the DEM approach is coupled with a numerical flow simulation (CFD). Therefore, the commercial solver ANSYS Fluent® (*Release 17.1*) was applied. In the following sections, first the modelling of the fluid phase, followed by the particle phase and finally the coupling method of both phases will be described.

3.1 Fluid phase modelling

3.1.1 Governing equations

In the current CFD approach, the computational domain is divided into control volumes CV or cells (figure 3.1), in which the spatial porosity distribution is defined.

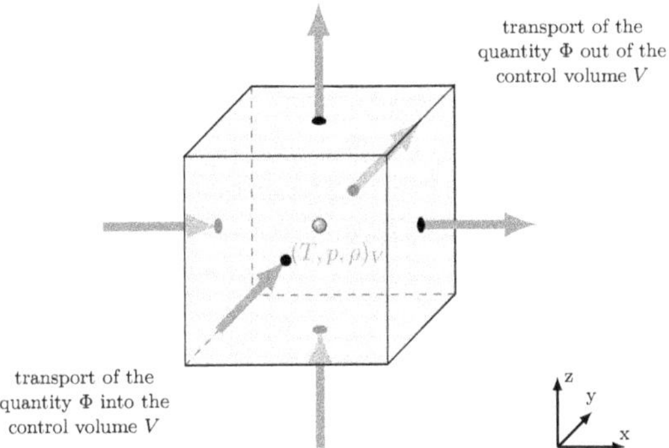

Figure 3.1: Stationary control volume V with the in- and outflow directions of the physical transport quantity Φ

Note that the porosity is calculated in the DEM (see subsection 3.3.2). For this control volume conservation equations for mass and momentum are derived. These and other scalar transport quantities are interpolated to the finite control volume and balanced per unit time. Here, the general physical quantity Φ enters or exits through the surfaces of the control volume.

In the following, the conservation equations for mass (equation (3.1)) and momentum (equation (3.3)) will briefly be discussed. A detailed derivation of these equations can be found in the work of Oertel [162].

Equation of continuity The mass conservation equation is also known as continuity equation. The change in mass over time within CV is calculated according to [1]

$$\frac{\partial\left(\varepsilon_f \rho_f\right)}{\partial t}+\nabla\left(\rho_f \boldsymbol{u}_{f,sup}\right)=0 \qquad [\mathrm{kg\,s^{-1}\,m^{-3}}] \tag{3.1}$$

with $\boldsymbol{u}_{f,sup}$, ρ_f and ε_f representing the superficial fluid velocity, fluid density and the local porosity, respectively.
The conversion from physical velocity $\boldsymbol{u}_{f,phys}$ to empty-pipe velocity $\boldsymbol{u}_{f,sup}$ is performed with

$$\boldsymbol{u}_{f,sup}=\varepsilon_f \boldsymbol{u}_{f,phys}. \qquad [\mathrm{m\,s^{-1}}] \tag{3.2}$$

Navier-Stokes equations The Navier-Stokes equations are based on the conservation equation of momentum. They read as [1]:

$$\begin{aligned}&\frac{\partial\left(\varepsilon_f \rho_f \boldsymbol{u}_{f,sup}\right)}{\partial t}+\nabla\left(\rho_f \boldsymbol{u}_{f,sup} \boldsymbol{u}_{f,sup}\right)\\&=-\varepsilon_f \nabla p+\nabla\left(\varepsilon_f \bar{\bar{\tau}}\right)+\varepsilon_f \rho_f \boldsymbol{g}+\boldsymbol{f}_{int} \qquad [\mathrm{N\,m^{-3}}]\end{aligned} \tag{3.3}$$

Here, $\boldsymbol{g}$ and ∇p refer to acceleration of gravity and pressure gradient, respectively. $\boldsymbol{f}_{int}$ is the source term which represents the exchange of momentum between solid and fluid phase. The left side of the equation combines the transient (time-dependent) and convective fluid acceleration. To the right, the first two terms represent the diffusion processes, either caused by pressure gradients or by frictional forces. In this context, $\bar{\bar{\tau}}$ denotes the fluid viscous stress tensor and can be calculated with the effective fluid viscosity η_f as follows:

$$\bar{\bar{\tau}}=\eta_f\left[\left(\nabla \boldsymbol{u}_{f,sup}\right)+\left(\nabla \boldsymbol{u}_{f,sup}\right)^{-1}\right] \qquad [\mathrm{N\,m^{-2}}] \tag{3.4}$$

The third term refers to the influences caused by gravity. When gas and solid particles are moving relative to each other, the resulting particle-fluid interaction is determined by the additional source term $\boldsymbol{f_{int}}$ in each CFD cell, which is given component-wise as follows

$$f_{int_j} = \bar{\beta}_j \left(u_{f,sup,j} - \bar{v}_j\right) \qquad [\mathrm{N\,m^{-3}}] \qquad (3.5)$$

where $\bar{v}_j$ is the fluid cell-averaged particle velocity and $\bar{\beta}_j$ is the fluid cell averaged particle-fluid friction coefficient. Both values are calculated within the DEM and have to be transferred to the fluid cells $j = x, y, z$ as time-averaged values due to the typically smaller time step needed in the Discrete Element Method compared with the CFD calculation. A detailed description of the coupling mechanism applied is given in section 3.3.3.

3.1.2 Turbulence modelling

In the current thesis, Reynolds-Averaged Navier-Stokes approaches (RANS) in form of two-equation models are applied for turbulence modelling. The two-equation models belong to the most frequently applied approaches for industrial tasks. These approaches are named after the two additional transport equations which are solved to cover the influence of turbulence on the fluid flow behaviour. Apart from the often applied Standard-k-ε model [21, 41, 117, 131], the k-ε-Realizable model [169, 176], the Low-Re-k-ε model [138] and the k-ω model [122] are also commonly used.

The Standard-k-ε model has already been used in many applications, so the limits of this model are well known. This model is commonly applied within the analysis of technical flow processes, where the global flow profile plays a major role. Especially central zones of a full-turbulent pipe flow can be modelled accurately. The weakness of the k-ε model is its limited predictability for complex flows with e.g. stagnation points, curved and/or twisted streamlines [199]. Therefore, in the current thesis the so-called Shear-Stress-Transport model, in short k-ω-SST model, is applied.

k-ω-SST model It is based on both two-equation models Standard-k-ε and k-ω. Therein the characteristic frequency or so-called specific dissipation rate ω of the energetic vortices with ε (dissipation rate) and k (specific turbulence kinetic energy) is considered as a turbulence variable:

$$\omega = \frac{\varepsilon}{k} \qquad [\mathrm{s^{-1}}] \qquad (3.6)$$

Details concerning the internal modelling equations are presented extensively by Wilcox [234].

As a hybrid approach, the k-ω-SST model applies two different equations for ω. Close to walls the k-ω model for calculating the average turbulence and flow variables is obtained here. In the outer zones, the ω-equation is determined by a transformation using the transport equation for ε, so that in these zones a k-ε model is effectively used for turbulence and flow simulation. Details about the modelling and the cross-fade between both equations can be taken from [64, 147, 148, 199]. The k-ω-SST model by Menter exploits the advantages of both the k-ω and the Standard k-ε model and thus compensates the disadvantages of both models, respectively.

3.2 Solid phase modelling

The current thesis uses the Discrete Element Method [44]. In the following, the basics of DEM are briefly discussed. Details on the mechanical and mathematical fundamentals of the DEM have been presented extensively in further literature, e.g. [90, 120].

3.2.1 Governing equations

In the DEM, a balance of forces is determined for each individual particle. An overview of the three relevant scenarios is given in figure 3.2.

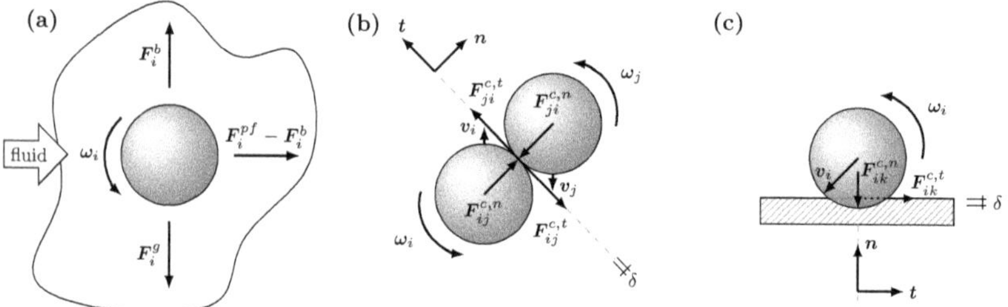

Figure 3.2: Balance of forces affecting a single particle within a flow field (a), during inter-particle contact (b) and at particle-wall collision (c)

Lagrangian equations of motion The translational particle motion and the particle rotation result from momentum conservation or the Newtonian (equation (3.7)) and Eulerian (equation (3.8)) relations:

$$m_i \cdot \frac{d\boldsymbol{v}_i}{dt} = \sum_{j=1}^{N} \boldsymbol{F}_{ij}^{c} + \boldsymbol{F}_i^{pf} + m_i \boldsymbol{g}, \qquad [\mathrm{N}] \quad (3.7)$$

$$\theta_i \cdot \frac{d\boldsymbol{\omega}_i}{dt} = \sum_{j=1}^{N} \boldsymbol{M}_{ij} + \boldsymbol{M}_i^{r} + \boldsymbol{M}_i^{pf} = \sum_{j=1}^{N} \boldsymbol{r}_i \times \boldsymbol{F}_{ij}^{c} + \boldsymbol{M}_i^{r} + \boldsymbol{M}_i^{pf}. \qquad [\mathrm{N\,m}] \quad (3.8)$$

In equation (3.7), m_i and $\boldsymbol{v}_i$ denote the particle mass and the particle's translational velocity, respectively. $\boldsymbol{F}_{ij}^{c}$ refers to the force acting on each particle due to the impact between the contact pair i and j, whereas $\boldsymbol{F}_i^{pf}$ and $m_i\boldsymbol{g}$ represent the additional force

impressed on the particle i by the surrounding fluid and the acceleration of gravity, respectively. In equation (3.8), θ_i is the mass tensor of inertia. $\boldsymbol{\omega}_i$, $\boldsymbol{r}_i$ and $\boldsymbol{M}_i^r$ refer to the direction-dependent angular velocity, the lever applied for the corresponding contact case and the torque acting due to rolling friction, respectively. The torque acting on a particle due to particle-fluid interaction is represented by $\boldsymbol{M}_i^{pf}$.
The determination of the translational $\boldsymbol{v}_i$ and the rotational velocity $\boldsymbol{\omega}_i$ is obtained by numerical time-integration of equations (3.7) and (3.8) with the forces and torques acting on the particle at each time step. Various integration schemes have already been extensively discussed in literature [119]. In the present work, the semi-implicit Euler-Cromer method is applied [43]. By determination of translational and angular velocities, the change in position and rotation of a single particle within a defined time step can be calculated.

3.2.2 Contact forces

Generally, it is possible to determine forces and plastic deformations caused by individual contacts in a time-resolved way, but doing so requires an enormous computational effort and is therefore only suitable for systems with small numbers of particles.
For particle systems with many particle-particle or particle-wall interactions, modelling elastic collisions of rigid bodies using a virtual overlap is more feasible by the so-called Soft-Sphere approach [44]. By separating the forces $\boldsymbol{F}_{ij}^{c}$ acting at the contact point into a normal component $\boldsymbol{F}_{ij}^{c,n}$ (in direction of the particle center of gravity $\boldsymbol{n}$) and a tangential part $\boldsymbol{F}_{ij}^{c,t}$ (in direction of the contact plane $\boldsymbol{t}$), both components can be numerically integrated independently of each other with specific contact models (equation (3.9)). Extensive overviews of approaches most frequently used within DEM and their respective modifications are given by Kruggel-Emden [118, 120] and Di Renzo and Di Maio [52, 53].

$$\boldsymbol{F}_{ij}^{c} = \boldsymbol{n} \cdot \boldsymbol{F}_{ij}^{c} + \boldsymbol{t} \cdot \boldsymbol{F}_{ij}^{c} = \boldsymbol{F}_{ij}^{c,n} + \boldsymbol{F}_{ij}^{c,t} \qquad [\mathrm{N}] \qquad (3.9)$$

Both $\boldsymbol{F}_{ij}^{c,n}$ and $\boldsymbol{F}_{ij}^{c,t}$ are determined separately using semi-empirical, so-called spring-dashpot models [120]. Figure 3.3 depicts the principle of the model applied for determining the normal (a) and tangential (b) part. Here, the principle is exemplified by particle-particle contacts, which can be transferred analogously to particle-wall collisions.

Normal force Within the viscoelastic spring-dashpot model applied, the normal force is split into an elastic $\boldsymbol{F}_{el}^{c,n}$ and a dissipative term $\boldsymbol{F}_{diss}^{c,n}$:

$$\boldsymbol{F}^{c,n} = \boldsymbol{F}_{el}^{c,n} + \boldsymbol{F}_{diss}^{c,n} = -k^n \delta \boldsymbol{n} - \gamma^n \boldsymbol{v}_{rel}^{n} \qquad [\mathrm{N}] \qquad (3.10)$$

While for the elastic term, a linear dependence between the virtual overlap δ and a spring stiffness k^n is assumed, the determination of the dissipative component is based on a damper, which is derived from the product of an empirically obtained damping coefficient γ^n and the relative particle velocity in normal direction $\boldsymbol{v}_{rel}^{n}$. The uniform normal vector faces along the line connecting both particles' centers of gravity ($\boldsymbol{x}_i$, $\boldsymbol{x}_j$) and can be determined to

$$\boldsymbol{n} = \frac{\boldsymbol{x}_j - \boldsymbol{x}_i}{|\boldsymbol{x}_j - \boldsymbol{x}_i|}. \qquad [-] \qquad (3.11)$$

The displacement of the spring represents the virtual overlap of both contact partners (see figure 3.3 (a)) and is determined from the distance between the (spherical) particle centres of gravity and the two radii. A contact (and thus an overlap) exists if the distance between both particle centres of gravity is smaller than the sum of their radii. However, the spring stiffness results from the material types of the interacting particles, the coefficient of restitution e_n and the collision time t_n [239]. To describe the contacts in DEM time-resolved, it is necessary to select an integration time step size that is much smaller than the contact duration, which primarily depends on the collision velocity.

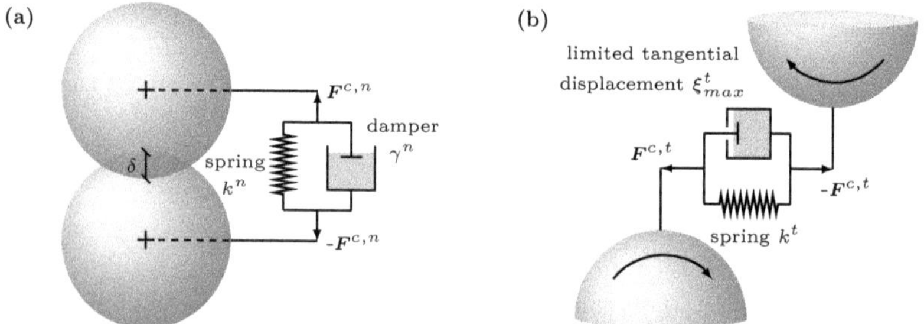

Figure 3.3: Linear spring-dashpot model for normal (a) and tangential force (b) description

Tangential force Following the Coulomb law, the tangential force according to equation (3.12) is directly dependent on the effective normal force $\boldsymbol{F}^{c,n}$. Here, a distinction is made between cases of static friction, where the relative tangential velocity $\boldsymbol{v}^t_{rel}$ between both contact partners is zero, and dynamic sliding friction, where the relative velocity is not zero.

$$\boldsymbol{F}^{c,t}_{Coul} = \begin{cases} \mu_{stat} \cdot |\boldsymbol{F}^{c,n}| \cdot \boldsymbol{t} & \text{for} \quad \boldsymbol{v}^t_{rel} = 0 \\ \mu_{dyn} \cdot |\boldsymbol{F}^{c,n}| \cdot \boldsymbol{t} & \text{for} \quad \boldsymbol{v}^t_{rel} \neq 0 \end{cases} \qquad [\text{N}] \qquad (3.12)$$

The coeffients for static friction μ_{stat} and for dynamic friction μ_{dyn} are in the same range for similar pairs of materials, which is why many DEM approaches often assume that $\mu_c \approx \mu_{stat} \approx \mu_{dyn}$ [12, 90, 203].

The linear tangential spring-dashpot model applied in the scope of this thesis is schematically depicted in figure 3.3 (b). The tangential force is modelled by a spring with the stiffness k^t and a parallel connected dashpot with a damping coefficient γ^t.

$$\boldsymbol{F}^{c,t}_{SD} = -k^t \boldsymbol{\xi}^t - \gamma^t \boldsymbol{v}^t_{rel} \qquad [\text{N}] \qquad (3.13)$$

The spring displacement $\boldsymbol{\xi}^t$ is initialised with the contact of two elements at contact time $t_0 = 0$ and is determined during the entire contact according to the tangential relative velocity.

$$\boldsymbol{\xi}^t = \left(\int_{t_0}^{t} \boldsymbol{v}_{rel}^t dt\right) \cdot \boldsymbol{t} \quad \text{with} \quad \boldsymbol{t} = \frac{\boldsymbol{v}_{rel}^t}{|\boldsymbol{v}_{rel}^t|} \qquad [\mathrm{m}] \qquad (3.14)$$

For ensuring the consistency of the tangential force, i.e. to avoid exceeding the Coulomb friction, the maximum spring displacement ξ_{max}^t is limited according to equation (3.15).

$$k^t \xi_{max}^t = \mu_c |\boldsymbol{F}^{c,n}| \qquad [\mathrm{N}] \qquad (3.15)$$

Thus, the resulting tangential force $\boldsymbol{F}^{c,t}$, as applied in the current approach, is obtained from the minimum of the Coulomb frictional force $\boldsymbol{F}_{Coul}^{c,t}$ and the calculated force $\boldsymbol{F}_{SD}^{c,t}$ using the spring-dashpot model.

$$\boldsymbol{F}^{c,t} = -\min\left(\boldsymbol{F}_{SD}^{c,t}, \boldsymbol{F}_{Coul}^{c,t}\right) = -\min\left(\left|-k^t \boldsymbol{\xi}^t - \gamma^t \boldsymbol{v}_{rel}^t\right|, |\mu_c \boldsymbol{F}^{c,n}|\right) \cdot \boldsymbol{t} \qquad [\mathrm{N}] \qquad (3.16)$$

Rolling friction $\boldsymbol{M}_i^r$ describes the energy dissipation of two contacting elements rotating relative to each other. An overview of rolling friction models can be found in the work of Ai et al. [5]. The rolling friction model applied in the current approach is based on the work of Beer and Johnston [13] and Zhou et al. [259]. The model describes a linear relationship between the torque due to rolling friction, a constant rolling friction coefficient μ_r and the product of the normal force, the effective rolling radius R and the vector of the rotational movement.

$$\boldsymbol{M}_i^r = -\mu_r \left|\boldsymbol{F}_{ij}^n\right| R \frac{\boldsymbol{\omega}_i}{|\boldsymbol{\omega}_i|} \quad \text{with} \quad \boldsymbol{\omega}_i = \frac{d\boldsymbol{\varphi}_i}{dt}, \quad R = \frac{r_i r_j}{r_i + r_j} \qquad [\mathrm{N\,m}] \qquad (3.17)$$

Here, $\boldsymbol{\varphi}_i$ and r_i or r_j represent the particle's orientation angles and the appropriate contacting particles' radii, respectively.

Parameter and coefficient determination The models described in the previous section for simulating macroscopic particle motion require an empirical determination of some model parameters:

- spring stiffnesses k^n, k^t (cf. equations (3.10) and (3.13))
- dashpot coefficients γ^n, γ^t (cf. equations (3.10) and (3.13))
- Coulomb friction μ_c (cf. equation (3.12))
- rolling friction μ_r (cf. equation (3.17))

These material-dependent parameters must be determined empirically for each potential material combination during a DEM simulation. The friction coefficients μ_c and μ_r can be determined by corresponding force measurements [90]. Since the spring stiffnesses k^n

and k^t and the dashpot coefficients γ^n and γ^t represent the elastic and plastic component of an impact only within the spring-dashpot model (dissipation of kinetic energy), these non-physical parameters are not directly measurable. The appropriate specification can be performed directly by measuring the coefficient of restitution e_n or the collision time t_n [194]. These quantities can be determined by corresponding impact experiments or drop tests [42, 58]. The data determined from such experiments only serve as rough approximations of the model parameters. In most cases, a further iterative adaption and verification by additional experiments, such as static or dynamic experiments of the angle of repose are required [180–182]. This procedure was performed for determination of all parameters required for the current case of wood pellet-DEM simulation. The appropriate material-dependent parameters and coefficients required are listed in table 5.2.

3.2.3 Particle shape

Due to the very simple contact detection between particles, spheres are often used as body approximations in DEM simulations. However, spheres cannot fully describe the behaviour (angle of repose, entanglement, etc.) of technically relevant non-spherical particle bulk materials [90].

Possible approaches are those which define the discrete objects for contact determination as spheres and adapt the force model to consider cohesion [20, 238] or simply modify the force parameters to mimic the motion of particles of complex shape [58]. In this thesis, a degree of resolution should be achieved, which allows the modelling of the shape-dependent motion and flow behaviour of cylindrical wood pellets. Although generic models for describing the mechanical behaviour of cylindrical particles already exist [62], these usually require complex methods for consideration of special cases in contact detection at the transition between shell and end face. Various body types (polyhedra, spherical clusters, superellipsoids) have been applied for mathematical approximation of cylindrical geometries in DEM simulations [77, 156, 190]. Höhner et al. [92], for example, demonstrated the general suitability of polyhedra based on triangulated surfaces for representing the motion behaviour of wood pellets discharging from a hopper. Nevertheless, the disadvantage of the approximation of non-spherical particles as polyhedra is the enormous computational capacity and time required for contact determination [24]. To avoid this effect using polyhedra, the current thesis follows a suggestion of Williams and Philipse [237] who approximate cylindrical particles as spherocylinders.

Figure 3.4 illustrates the geometric definition of the pellet approximation as spherocylinders and the overlap δ to be determined in case of contact. A spherocylinder can be completely defined by its radius r_p, its length l_p and a rotational axis $\boldsymbol{RA}$. The length of $\boldsymbol{RA}$ of the ideal cylindrical surface is calculated by equation (3.18). Since the shortest distance between a contact point on the rotational axis and the corresponding surface of the spherocylinder is always represented by the radius, the contact point and the resulting overlap can be determined for each collision case using both the radii of the contacting spherocylinders and this distance.

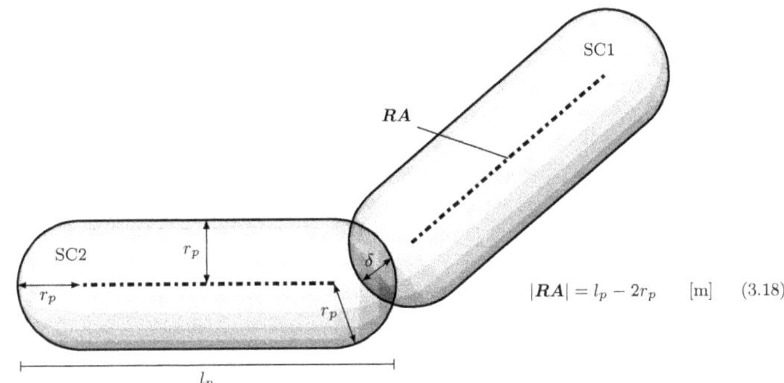

$$|\boldsymbol{RA}| = l_p - 2r_p \qquad [\mathrm{m}] \qquad (3.18)$$

Figure 3.4: Schematic representation of two touching spherocylinders and definition of the rotational axis $\boldsymbol{RA}$

3.2.4 Neighbour and contact detection

For an efficient determination of particle contacts and overlaps relevant for mechanics determination, most DEM approaches perform a neighbour detection before the actual contact determination (see figure 3.5).

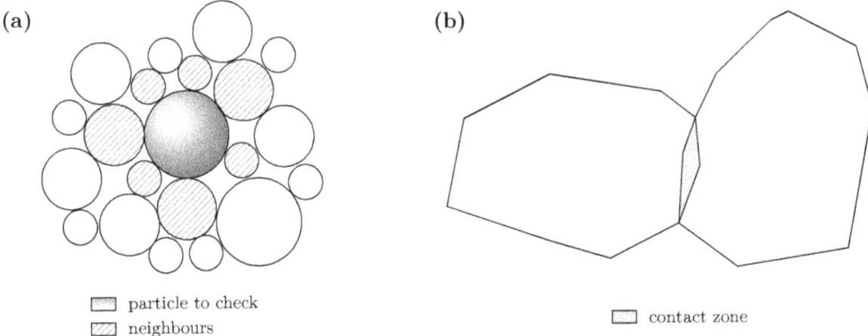

Figure 3.5: Division of the contact determination routine into neighbour detection (a) and geometric contact resolution (b) [90]

The aim of these detection algorithms is to generate lists of possible neighbours as short as possible for checking respective contacts. Therefore, as many possible contact pairs should be excluded in the neighbourhood detection algorithm. Depending on the application (particle size distribution, particle geometry and system dynamics), different neighbour detection algorithms have been established, such as grid-based methods [235], tree-based algorithms [61] or Delaunay triangulations [63]. Benchmark studies of the respective methods with respect to their computational efficiency are published e.g. in [81, 163]. Within the DEM approach applied in this thesis, a contact determination algorithm was chosen which is based on the so-called *Contact Grid* (CGRID) algorithm developed by Williams et al. [235].

Figure 3.6 illustrates the principle of the CGRID algorithm applied with application to spherocylinders. Here, the cells of the underlying contact grid are shown. The dashed, rectangular frames define the zone where a spherocylinder could be located. The gray hatched zones mark those cells of the underlying grid that are in contact with the dashed bounding box and thus potentially contain segments of the spherocylinders (so-called *mapping*).

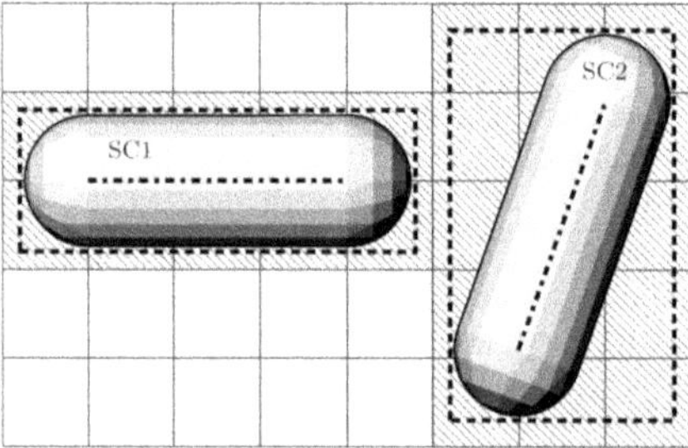

Figure 3.6: Working principle of CGRID neighbourhood detection

At the beginning of a DEM simulation, the domain is divided into rectangular cells of the same size for each direction. This allows efficient indexing of the individual cells. Unlike tree-based approaches or Delaunay triangulations, these structures only need to be initialised once at the beginning of a simulation. The size of the grid cells does not necessarily have to be larger than the particle size, which leads to a possible simultaneous association of a particle to several grid cells. In his thesis, Höhner [90] determined that an approximate cell size of 1.5 times of the surrounding sphere diameter of the smallest particle roughly provides the optimum regarding computational efficiency.

The particle is associated to the occupied cells of the underlying grid using the minima and maxima of the particle's bounding box in each direction (see figure 3.6). Since particles can be associated to several grid cells, it is possible that for certain contact pairs the contact determination routine is performed several times. Therefore, the contact pair is attached only once to the contact list.

Neighbourhood detection with static underlying grids allow the definition of cuboid borders according to the maximum dimensions of the particle in the respective directions (see figure 3.6). The scenarios illustrated in figure 3.6 depict the most favourable (SC1) and a rather unfavourable case (SC2) of the particle orientation in a Cartesian, two-dimensional coordinate system for an efficient contact determination with the CGRID algorithm.

The contact determination itself has been performed by the so-called Fast Common Plane algorithm according to Nezami et al. [157]. Therein, the contact detection is simplified to two separate contact detection tests between each contact partner and a plane (Common Plane - CP) positioned between them. Before the contact area of both partners can be approximated, the algorithm determines the correct positioning specified by defined conditions. These and details of the working principle are extensively presented by Höhner et al. [91, 93].

3.2.5 Coordinate systems and particle orientation

The motion of non-spherical particles within fluid flows depends on their shape and orientation. For describing the particle orientation, a global coordinate system of the simulation environment $[x, y, z]$ and a local particle system $[x', y', z']$ with origin in the geometric particle center is defined in analogy to Yin et al. [246, 247] (cf. figure 3.7 (a)).

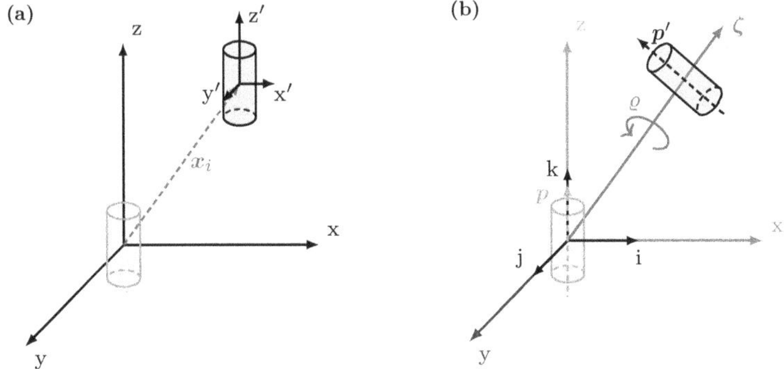

Figure 3.7: Definition of coordinate systems and particle translation (a) and particle rotation via quaternion (b)

Quaternions [80, 125] are employed in the current DEM approach to represent particle orientation in space as well as rotations. A quaternion builds a four-components vector, whereby three of its elements can be considered as a separate three-dimensional vector. These three components define a normalised axis, whereas the fourth describes the angle of rotation around this axis.

Compared to the description of rotations by Euler angles, quaternions have the advantage to being independent on the sequence of rotations around the spatial axes, which can lead to the so-called gimbal lock effect when using Euler angles. By considering three independent axes of rotation, it can happen that two of them are aligned parallel to each other if one of them is exactly 90° [231]. A rotation around the other two rotational axes then leads to the same rotation, so that rotation around a spatial axis is blocked.

In addition, rotation matrices have limitations related to memory, CPU load, numerical errors and overdetermined data [125].

Quaternions are basically applied for describing orientations and rotations. A quaternion $\underline{q}$ is defined by i, j, k as three imaginary bases and q_0, q_1, q_2, q_3 as scalars as follows:

$$\underline{q} = q_0 + q_1 i + q_2 j + q_3 k \qquad [-] \qquad (3.19)$$

Related to the imaginary range, q_0 represents the real part and q_1, q_2, q_3 the imaginary parts, respectively.
The imaginary bases and the real base build a four-dimensional vector space. For quaternions, the associative and distributive law as well as the multiplicative invertibility apply, but not the commutative law. For the imaginary bases the following Hamilton laws apply [72]:

$$i^2 = j^2 = k^2 = ijk = -1 \qquad [-] \qquad (3.20)$$

The rotation of a vector $\boldsymbol{p}$ by a quaternion $\underline{q}$ to a resulting vector $\boldsymbol{p}'$ is performed by multiplying $\underline{q}$, $\boldsymbol{p}$ and the inverse of the quaternion $\underline{q}^{-1}$:

$$\boldsymbol{p}' = \underline{q} \cdot \boldsymbol{p} \cdot \underline{q}^{-1} \qquad [-] \qquad (3.21)$$

The inverse quaternion is calculated with regards to equation (3.19) by

$$\underline{q}^{-1} = \frac{q_0 - q_1 i - q_2 j - q_3 k}{q_0^2 + q_1^2 + q_2^2 + q_3^2}. \qquad [-] \qquad (3.22)$$

The length of a quaternion is defined analogous to the Euclidean standard of a three-dimensional vector and can be used to define a unit quaternion. Since the uniform quaternion has a length of $q_0^2 + q_1^2 + q_2^2 + q_3^2 = 1$, equation (3.22) is directly transformed to

$$\underline{q}^{-1} = q_0 - q_1 i - q_2 j - q_3 k. \qquad [-] \qquad (3.23)$$

A quaternion is finally defined by a normalised vector $\boldsymbol{\zeta}$ and the rotation angle ϱ around it

$$\underline{q} = q_0 + q_1 i + q_2 j + q_3 k = \cos\left(\frac{\varrho}{2}\right) + \sin\left(\frac{\varrho}{2}\right)\boldsymbol{\zeta}. \qquad [-] \qquad (3.24)$$

To describe the rotation of an object or vector via quaternions, the rotated vector $\boldsymbol{p}'$ and the initial vector $\boldsymbol{p}$ must be known. If the initial vector $\boldsymbol{p}_n$ and the current vector $\boldsymbol{p}_n{}'$ can be defined as normalised vectors respectively (e.g. axis of a cylinder, see figure 3.7 (b)), the rotated quaternion can be represented by the vector $\boldsymbol{\zeta}$ and the angle ϱ, which are calculated as follows:

$$\boldsymbol{\zeta} = \begin{pmatrix} \zeta_x \\ \zeta_y \\ \zeta_z \end{pmatrix} = \boldsymbol{p}_n \times \boldsymbol{p}_n{}' \qquad [-] \qquad (3.25)$$

$$\varrho = \arccos\left(\boldsymbol{p}_n \cdot \boldsymbol{p}_n{}'\right) \qquad [°] \qquad (3.26)$$

Vector $\boldsymbol{\zeta}$ results from the cross product of the initial vector and the rotated one. The angle ϱ is derived from the angle between both vectors. Figure 3.7 (b) depicts schematically the relation of the described parameters. Vector $\boldsymbol{p}$ represents the initial orientation of a cylindrical object using the rotational axis through the center of gravity, vector $\boldsymbol{p}'$ describes the new particle orientation resulting from the rotation of the object around $\boldsymbol{\zeta}$ with the angle ϱ. Note that due to visualisation reasons the rotated particle is additionally translated. Thus, the position and orientation of an object can be described completely and without further assumptions by center of gravity $\boldsymbol{x}_i$ and the appropriate quaternion $\underline{q}$.

3.3 Modelling of particle-fluid interaction

In the following, the models employed to determine particle-fluid interaction forces and torques related to cylindrical particle geometry are introduced first. Next, the porosity determination is discussed before the implemented coupling mechanisms for continuous data transfer between the in-house DEM code and the CFD solver ANSYS Fluent® is explained in detail. Finally, the parameters to be transferred between both simulation routines are discussed more specifically.

3.3.1 Particle-fluid interaction

The different interaction forces acting on a cylindrical particle within fluid flow are depicted with their appropriate points of action (i.e. geometric center and center of pressure) in figure 3.8.

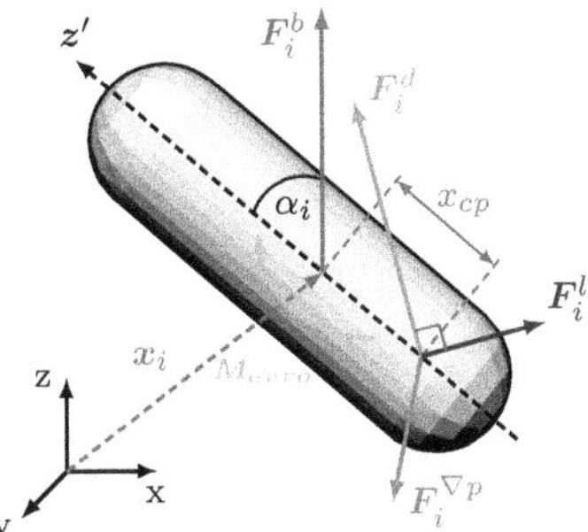

Figure 3.8: Particle-fluid interaction forces and torques acting on a particle with their point of application

The predominant particle-fluid interaction forces $\boldsymbol{F}_i^{pf}$ can be summed up as:

$$\boldsymbol{F}_i^{pf} = \boldsymbol{F}_i^b + \boldsymbol{F}_i^d + \boldsymbol{F}_i^{\nabla p} + \boldsymbol{F}_i^l + \underbrace{\boldsymbol{F}_i^{vm} + \boldsymbol{F}_i^{Basset} + \boldsymbol{F}_i^{\nabla \bar{\bar{\tau}}} + \boldsymbol{F}_i^{other}}_{\approx 0} \qquad [\mathrm{N}] \qquad (3.27)$$

The buoyancy force $\boldsymbol{F}_i^b$ applies in the geometric center $\boldsymbol{x}_i$ of the particle and principally counteracts the gravity force in equation (3.7). Forces such as drag force $\boldsymbol{F}_i^d$, pressure gradient force $\boldsymbol{F}_i^{\nabla p}$ and lift force $\boldsymbol{F}_i^l$ act at the particle's center of pressure. Depending on the particle's orientation α_i, the center of pressure is located along the major particle axis $\boldsymbol{p}'$ displaced from the geometric center by the lever x_{cp}. Due to the point of application being offset from the geometric center, this results in the torque $\boldsymbol{M}_{aero}$. In the current work only gas-solid flows with small density ratios ($\rho_f \ll \rho_p$) are considered, so that the virtual mass force $\boldsymbol{F}_i^{vm}$ and the Basset history force $\boldsymbol{F}_i^{Basset}$ can be neglected [212, 249]. The virtual mass force is an unsteady force representing the acceleration of the surrounding fluid and the Basset force (also called *Basset history term*) accounts for viscous effects. Typically, the forces become insignificant for $\rho_f/\rho_p \leq 10^{-3}$ [205]. Hilton et al. [87, 88] show that under the assumption of a parallel turbulent flow the particle-fluid stress term in general can be neglected leading to $\boldsymbol{F}_i^{\nabla \bar{\bar{\tau}}} = 0$. In the approach applied, some other terms such as thermophoretic and Brownian force are also neglected and are cumulated in equation (3.27) to $\boldsymbol{F}_i^{other}$.

Buoyancy force The buoyancy force opposing the gravitational field results for a particle i to

$$\boldsymbol{F}_i^b = \rho_f \boldsymbol{g} V_i. \qquad [\mathrm{N}] \qquad (3.28)$$

Here, V_i and ρ_f represent the particle volume and the fluid density, respectively.

Drag and pressure gradient force According to Shimizu [201] and Bouillard et al. [17] the drag force and the pressure gradient force acting on a single particle within a cell with the porosity ε_f can be combined to

$$\boldsymbol{F}_i^d + \boldsymbol{F}_i^{\nabla p} = \beta_{i,j} V_i \cdot \frac{\boldsymbol{u}_{f,phys} - \boldsymbol{v}_i}{\varepsilon_f (1 - \varepsilon_f)}. \qquad [\mathrm{N}] \qquad (3.29)$$

A detailed derivation is given by Hold [94]. For the determination of the particle-fluid friction coefficient $\beta_{i,j}$ a large number of models that are based on empirically and/or numerically determined correlations exist in literature. A comprehensive description of various models can be found in the work of Gidaspow [70], Zhu et al. [262], Theidel [212], Hold [94] and Krüger [116]. Very popular and widely used for spherical but also applicable for non-spherical particles for describing multiphase systems [251] is the approach of Di

Felice [51]. Therein, the interacting force on an isolated particle is calculated and altered by the influence of surrounding particles, regarded by the porosity function $\varepsilon_f^{(1-\chi)}$:

$$\boldsymbol{F}_i^d + \boldsymbol{F}_i^{\nabla p} = \frac{1}{2}\rho_f C_D A_\perp |\boldsymbol{u}_{f,phys} - \boldsymbol{v}_i| \left(\boldsymbol{u}_{f,phys} - \boldsymbol{v}_i\right) \varepsilon_f^{(1-\chi)} \qquad [\mathrm{N}] \qquad (3.30)$$

Here, C_D is the drag coefficient and $A_\perp$ refers to the cross-sectional area perpendicular to the flow. The exponential correction factor χ is given by

$$\chi = 3.7 - 0.65 \exp\left(-\left(1.5 - \log\left(Re_p\right)\right)^2/2\right). \qquad [-] \qquad (3.31)$$

χ is calculated as a function of the particle's Reynolds number Re_p

$$Re_p = \varepsilon_f \rho_f l_{char} \cdot \frac{|\boldsymbol{u}_{f,phys} - \boldsymbol{v}_i|}{\eta_f} \qquad [-] \qquad (3.32)$$

with the particle's characteristic length l_{char}, which is commonly assumed as the diameter of the volume equivalent sphere of the appropriate particle and the dynamic viscosity of the fluid η_f.

Regarding equation (3.29), equation (3.30) can be rewritten in terms of the particle-fluid friction coefficient as

$$\beta_{i,j} = \frac{1}{2}\rho_f C_D A_\perp \varepsilon_f |\boldsymbol{u}_{f,phys} - \boldsymbol{v}_i| \left(1 - \varepsilon_f\right) \frac{1}{V_i} \varepsilon_f^{(1-\chi)}. \qquad [-] \qquad (3.33)$$

Drag coefficient The knowledge of the drag coefficient C_D is essential to model particle motion inside fluid flow. For a single particle (spherical or non-spherical), it can be derived from numerical detailed resolved flow simulations [96, 177, 249] or from empirical correlations. Various correlations have been established in the context of pneumatic conveying processes [94, 212], e.g. the approaches of Dalla Valle [45], Schiller and Naumann [196] or Hölzer and Sommerfeld [95]. The correlations of both Dalla Valle (equation (3.34)) or Schiller and Naumann (equation (3.35)) take only the particle Reynolds number but not the orientation into account. Thus, they are mostly suitable for spherical particles.

$$C_D = \left(0.63 + \frac{4.8}{\sqrt{Re_p}}\right)^2 \qquad [-] \qquad (3.34)$$

$$C_D = \begin{cases} \frac{24}{Re_p}\left(1.0 + 0.15 Re_p^{0.687}\right) & Re_p \leq 1000 \\ 0.44 & Re_p > 1000 \end{cases} \qquad [-] \qquad (3.35)$$

Due to its simplicity and general applicability concerning particle-fluid interaction of non-spherical particles, the model by Hölzer and Sommerfeld [95] is widely used [158, 253] and thus applied in the current approach. It can be written as

$$C_D = \frac{8}{Re_p}\frac{1}{\sqrt{\phi_{\parallel}}} + \frac{16}{Re_p}\frac{1}{\sqrt{\phi}} + \frac{3}{\sqrt{Re_p}}\frac{1}{\phi^{3/4}} + 0.42 \times 10^{0.4(-\log(\phi))^{0.2}}\frac{1}{\phi_{\perp}}. \qquad [-] \qquad (3.36)$$

The sphericity ϕ represents the ratio between the surface area of the volume equivalent sphere $A^{sph_{eq}}$ and that of the considered particle A^p. The crosswise sphericity $\phi_{\perp}$ is the ratio between the cross-sectional area of the volume equivalent sphere and the projected cross-sectional area of the considered particle $A^p_{\perp}$ and the lengthwise sphericity $\phi_{\parallel}$ denotes the ratio between the cross-sectional area of the volume equivalent sphere and the difference between half the surface area and the mean projected longitudinal cross-sectional area $\bar{A}_{\parallel}$ of the considered particle. Note that depending on the particle shape, both lengthwise and crosswise sphericity can be greater than one. They are derived from

$$\phi = \frac{A^{sph_{eq}}}{A^p}, \quad \phi_{\perp} = \frac{A^{sph_{eq}}_{\perp}}{A^p_{\perp}}, \quad \phi_{\parallel} = \frac{A^{sph_{eq}}_{\perp}}{0.5A^p - \bar{A}_{\parallel}}. \qquad [-] \qquad (3.37)$$

With regard to cylindrical particles, the projected cross-sectional areas to be determined are sketched schematically in the following figure 3.9.

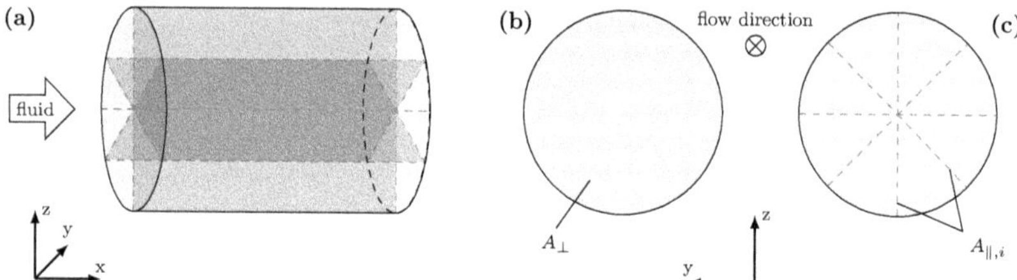

Figure 3.9: Overview of the flow-dependent areas needed for drag coefficient calculation by Hölzer and Sommerfeld [95] (a), projected cross-sectional area $A_{\perp}$ (b) and projected parallel areas $A_{\parallel,i}$ (c)

Due to computational reasons both the fluid flow-related cross-sectional and the mean projected longitudinal cross-sectional area are calculated in advance and are tabulated in dependence of the particle orientation.

To demonstrate the difference between the previously mentioned drag correlations (equations (3.34) to (3.36)) with respect to various particle shapes and orientations in the fluid flow, the drag coefficients are provided in figure 3.10 as a function of the turbulence regime or the particle Reynolds number.

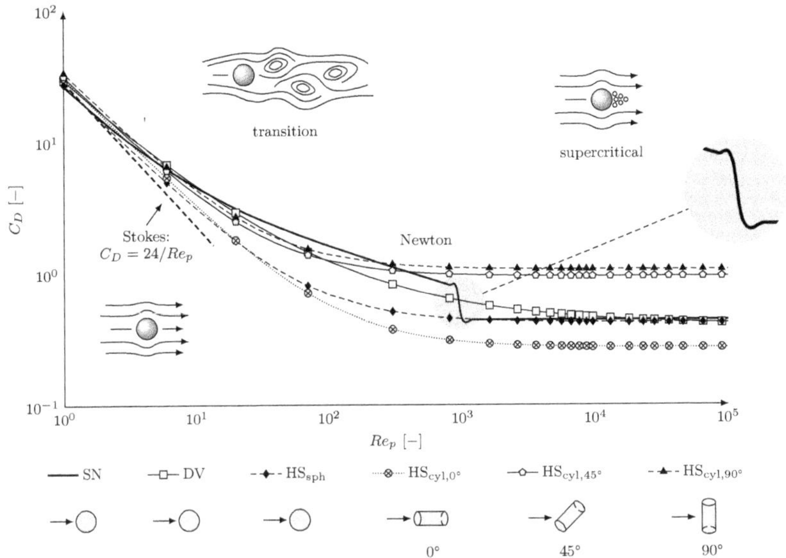

Figure 3.10: Dependence of the drag coefficient C_D on the particle Reynolds number Re_p regarding the correlations of Schiller and Naumann (SN), Dalla Valle (DV) or Hölzer and Sommerfeld (HS) for different particle shapes or orientations perpendicular to the flow

Note that the curves plotted in figure 3.10 apply for cylindrical particles with a diameter of 6 mm and a length of 18 mm. The volume-equivalent sphere applied has a diameter of approx. 10 mm.

The three models can be directly compared in case of flow-surrounded spheres, since the orientation can be neglected in these cases. The largest differences become obvious in the transition from Newtonian to supercritical turbulence flow regime ($Re_p \approx 10^3$). Among the three correlations considered, only Schiller-Naumann considers the flow separation with the Reynolds number-dependent case distinction (equation (3.35)). With lower Reynolds numbers, especially the Hölzer-Sommerfeld fit differs from both others and generally assumes significantly lower values for C_D. Starting at a Reynolds number of 10^3, the Schiller-Naumann and the Hölzer-Sommerfeld fits adopt almost identical values, whereas the Dalla Valle fit (previously rather an average value between both others) only approaches the other curves at Reynolds numbers higher than 10^4.

The dependence on particle shape and orientation is depicted by the Hölzer-Sommerfeld fits for variably arranged cylindrical particles. The values of C_D vary according to the cross-sectional area perpendicular to the flow, but are almost constant for Reynolds numbers lager than 10^3, which is the region of the supercritical flow regime. Reynolds numbers above the Newtonian regime occur mainly in pneumatic conveying processes with dilute phase flow regimes and are therefore highly relevant in the scope of the present work.

Lift force Basically, a lift force acting on a particle can be directly related to the drag force. In the current approach, the lift formation for cylindrical particles after Yin et al. [247] is applied, in which the normalised particle major axis ($\boldsymbol{p}_n{}'$) is used to characterise the force. In the current contribution, this force is only calculated for cylindrical particles and is neglected for spherical particles (fines), following Theidel [212]. It includes both the Saffman and the Magnus lift force. Considering that a) the lift force is orthogonal to the relative particle velocity ($\boldsymbol{u}_{f,phys} - \boldsymbol{v}_i$), b) lies in the plane defined by the particle's major axis direction and c) that the lift must be invariant under a 180°-rotation of the particle major axis $\boldsymbol{p}_n{}'$, i.e. vanishes if $\alpha_i = 0$ or π, the lift force is expressed as

$$\boldsymbol{F}_i^l = \frac{1}{2} C_L \rho_f \bar{A}_{\parallel} \frac{\boldsymbol{p}_n{}' \cdot (\boldsymbol{u}_{f,phys} - \boldsymbol{v}_i)}{|\boldsymbol{u}_{f,phys} - \boldsymbol{v}_i|} \left[\boldsymbol{p}_n{}' \times (\boldsymbol{u}_{f,phys} - \boldsymbol{v}_i)\right] \times (\boldsymbol{u}_{f,phys} - \boldsymbol{v}_i), \quad [\mathrm{N}] \qquad (3.38)$$

where C_L refers to the lift coefficient. Here, the usual assumption has been made that the lift is proportional to the drag [137] and that the dependence on the orientation is given by the so-called *cross-flow principle* with reference to Hoerner [89]:

$$\frac{C_L}{C_D} = \sin^2 \alpha_i \cdot \cos \alpha_i \qquad [-] \qquad (3.39)$$

For Reynolds-numbers in the Newtonian law regime ($30 < Re_p < 1500$) Mandø et al. [137] provide a more accurate correlation of the cross-flow principle:

$$\frac{C_L}{C_D} = \frac{\sin^2 \alpha_i \cdot \cos \alpha_i}{0.65 + 40 Re_p^{0.72}} \qquad [-] \qquad (3.40)$$

Torques For non-spherical (cylindrical) particles the centre of pressure does not coincide with the centre of mass at non-zero incidence angles, so that the aerodynamic forces mentioned before, which act at the centre of pressure rather than at the centre of mass, induce torques on the particle. Those can be split into torques due to aerodynamic forces and due to resistance. There exist different correlations depending on particle shape and orientation [137]. For cylindrical particles Yin et al. [247] introduced a correlation based on the work of Hoerner [89] taking the incidence angle α_i and the particle aspect ratio $\gamma = {}^{a}/_{b}$ into account. Herein, a and b refer to the cylinder radius and the half of the particle length, respectively. The authors determined the distance between the centre of pressure and the centre of mass for a cylinder as

$$x_{cp} = 0.25b \left(1 - e^{3(1-\gamma)}\right) \left|cos^3 \alpha_i\right|. \qquad [\mathrm{m}] \qquad (3.41)$$

Considering this lever, the torque due to the acting aerodynamic forces can be expressed as

$$\boldsymbol{M}_{aero} = (x_{cp} \cdot \boldsymbol{p}') \times \left(\boldsymbol{F}_i^d + \boldsymbol{F}_i^{\nabla p} + \boldsymbol{F}_i^l\right). \qquad [\mathrm{N\,m}] \qquad (3.42)$$

Since this torque is expressed in the inertial frame, it should be transformed to that in the particle frame, using the appropriate quaternion $\underline{q}_i$

$$\boldsymbol{M}'_{aero} = \underline{q}_i \cdot \boldsymbol{M}_{aero} \cdot \underline{q}_i^{-1}. \qquad [\mathrm{N\,m}] \qquad (3.43)$$

If a particle has an angular velocity $\boldsymbol{\omega}_i$ with respect to an axis, it will induce a torque of resistance on the rotating body. This torque will always act to reduce the angular velocity. After Yin et al. [247], for cylinders in a simple non-uniform flow field, this torque can be simply calculated for the respective components with

$$\begin{aligned} M'_{res,x} &= \frac{1}{64} C_D \rho_f \cdot 2a \cdot \omega_x^2 \cdot (2b)^4, & [\mathrm{N\,m}] \\ M'_{res,y} &= \frac{1}{64} C_D \rho_f \cdot 2a \cdot \omega_y^2 \cdot (2b)^4, & [\mathrm{N\,m}] \qquad (3.44) \\ M'_{res,z} &= \frac{1}{64} C_D \rho_f \cdot 2b \cdot \omega_z^2 \cdot (2a)^4. & [\mathrm{N\,m}] \end{aligned}$$

In contrast to the torque due to aerodynamic forces, this torque is defined directly in the particle frame. In the current approach, additional torques due to vorticity or shear inside the flow field are neglected, following [137, 247].
Thus, for cylindrical particles, both calculated torques due to particle-fluid interaction are summed up to

$$\boldsymbol{M}_i^{pf} = \boldsymbol{M}'_{aero} + \boldsymbol{M}'_{res}. \qquad [\mathrm{N\,m}] \qquad (3.45)$$

3.3.2 Determination of porosity

In the CFD framework, the continuous fluid domain is discretised into separate cells. The Lagrangian DEM particles must be mapped in the corresponding cells of the CFD mesh. Basically, two approaches exist (shown in figure 3.11) for mapping the DEM particles in the CFD mesh:

- **large (or coarse) grid** DEM-CFD approach
- **small (or fine) grid** DEM-CFD approach

The latter one is the more accurate method, but requires more computational resources. In contrast, the large grid DEM-CFD method is suitable for providing solutions for two-phase flows in realistic time scales. However, for modelling the particle-fluid interactions especially in the bend region precisely, a fine CFD mesh (small grid method) is applied in the current contribution.

According to equation (3.33), the calculation of the interacting forces is directly dependent on the locally and temporally changing solid fraction that is located within the considered control volume. In order to consider the cell volume blocked by the particle, the mean porosity (also void fraction) for each CV or cell is determined within DEM on the basis of the CFD mesh according to equation (3.46).

$$\varepsilon_{CV} = 1 - \frac{V_{solid,CV}}{V_{CV}} \quad \text{with} \quad V_{solid,CV} = \frac{V_{p,i}}{n_{CV,p}} \qquad [-] \qquad (3.46)$$

Figure 3.11 (in particular (b) and (c)) illustrates the methodology, where the occupied cell volume is determined by the sum of the partial volumes of the particles located in that cell. The volume of a particle $V_{p,i}$ is distributed proportionally among the number of control volumes $n_{CV,p}$ in which it is located.

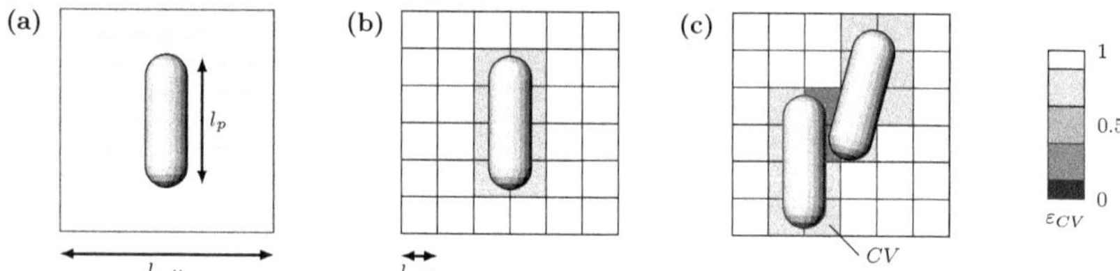

Figure 3.11: Principle of porosity determination including particles and voidage maps: coarse CFD grid with $l_{cell} \gg l_p$ (a), fine CFD grid with $l_{cell} < l_p$ (b) and CFD cells affected by multiple particles (c)

The fine grid method avoids overestimating boundary cells containing only a small fraction of particles, which would otherwise have been the case for the large grid approach. Compared to weakly loaded two-phase flows, the coarse grid method is more suitable for densely packed, slowly moving bulks as in blast furnaces [12].

3.3.3 Coupling implementation

Figure 3.12 depicts the coupling scheme of the coupled DEM-CFD approach. Before both simulations can be started, a stationary initial solution without source terms (and particles) is determined on CFD side and the particles are initialised in the DEM. In each time step, the DEM extracts the arithmetically averaged velocity vector of the particles, the vector of the averaged fluid friction coefficient and the void fraction for each CFD cell. The parameters are loaded into the CFD solver by a user-defined function (udf). A source term is determined subsequently. In turn, the velocity vector, density and dynamic viscosity of the fluid are written into a file via udf. Since the current simulations are transient, both methods always wait until new data are written respectively, so that the individual simulations run synchronously and the two individual solvers do not diverge. For example, the integration of the equation of motion (DEM) requires small time steps, whereas larger time steps suffice to resolve the fluid phase (CFD). The number of iterations or calculation time is used as termination criterion.

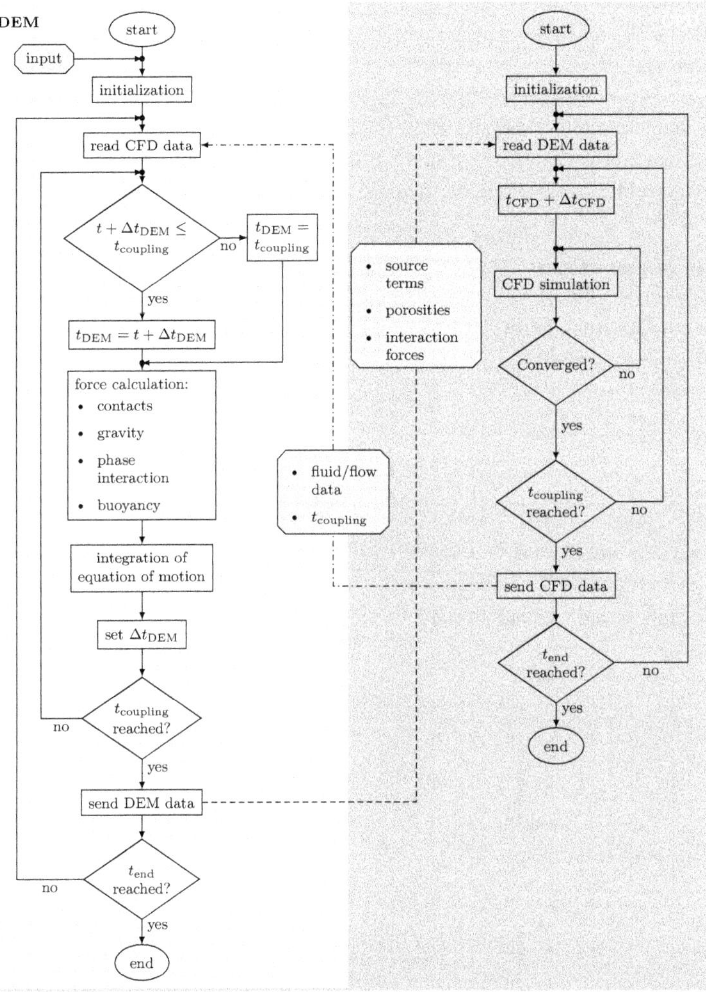

Figure 3.12: Sequence control of the DEM-CFD coupling method

For determination of the interaction parameters between both phases, the presented concept requires a continuous exchange of the coupling parameters. Thus, according to equation (3.5), the influence of the particles on the flow field in form of a source term is considered in the conservation equation of momentum. As mentioned in section 3.3.1, the fluid forces acting on a particle are determined on the basis of the local boundary conditions of the flow field. This requires a continuous bi-directional data transfer during runtime between both tools applied.

An explicit geometric assignment between solid and fluid phase is also required in order to exchange the local boundary conditions for the discrete particles as well as local sources and sinks for the conservation equations of the gas phase. While the particles are localised via their center of gravity, the properties of the fluid phase are assigned to the stationary cells of the discretised domain after solving the conservation equations. Thus, the coupling of both simulation tools demands in the first step a spatial assignment between CFD cells and particles and in the following step a calculation of the momentum source terms and the corresponding forces acting on the particle.

Transfer of momentum Figure 3.13 illustrates the data exchange between the DEM and CFD. Starting from the discrete single particle, the CFD cells are assigned to the particle in which it is completely or only partially located. The data $\Gamma_{i,j}$, necessary for the calculation of the interaction parameters, such as the flow velocity and the properties of the fluid phase, are averaged over all assigned cells n_i to specify the boundary conditions for the particle i.

$$\bar{\Gamma}_i = \sum \frac{\Gamma_{i,j}}{n_i} \qquad \text{[var.]} \qquad (3.47)$$

Based on the averaged fluid data, the calculation of the individual force components acting on the particle and the calculation of the cell-dependent averaged friction coefficient $\bar{\beta}_{CV}$ is carried out according to section 3.3.1.

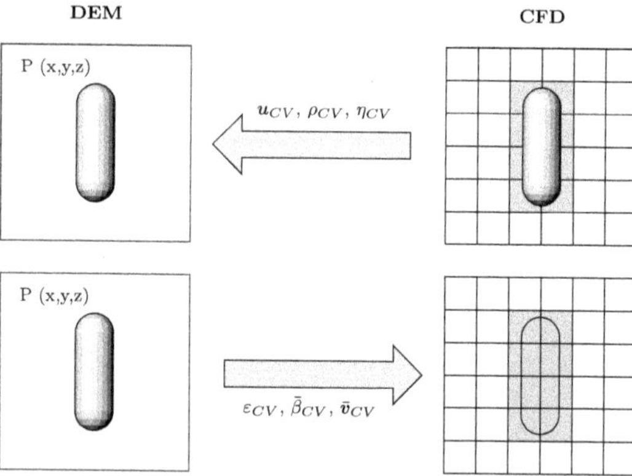

Figure 3.13: Working principle of momentum transfer between DEM and CFD [12]

Taking into account all particles or parts of them in one CFD cell, an average particle velocity $\bar{\boldsymbol{v}}_{CV}$ is determined. The data obtained and the previously determined cell porosity ε_{CV} are transferred to the CFD solver and used to determine the momentum source term according to equation (3.5).

4 Single particle degradation analysis

To develop empirical correlations for subsequent numerical prediction of pellet degradation characteristics, extensive experiments were performed using a single particle impact test facility, developed and adapted for cylindrical particle shape. After presenting the test facility, experimental results and the resulting correlations are discussed in the following. Afterwards, the implementation of the developed numerical model into the in-house DEM code is explained and verified.

4.1 Experimental setup

Design and operating principle The working principle of the single particle impact test facility is sketched in figure 4.1. Up to 100 pellets (maximum length and diameter of 50 and 6 mm, respectively) can be stored separately in the vertical bores of the rotating pellet depot. By rotating the depot bore by bore, each single pellet drops through a short tube into the rotation unit mounted underneath (figure 4.1, green circle), which rotates the pellet from vertical to horizontal position (shaft diameter 50 mm, bore diameter 7 mm).

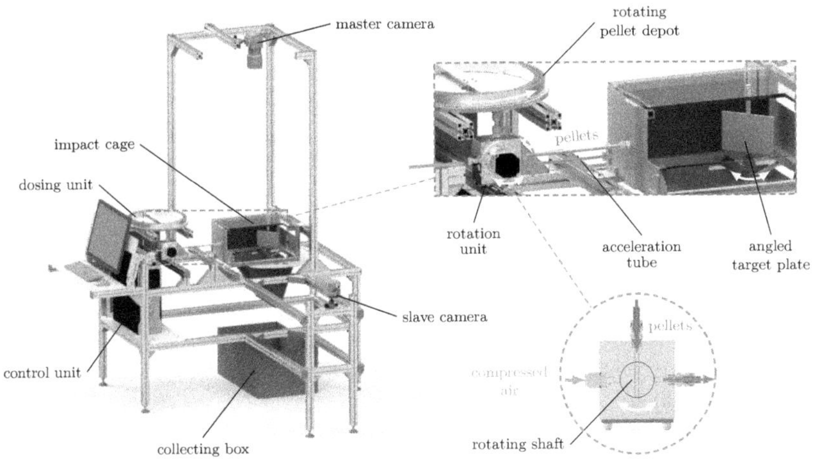

Figure 4.1: Design and working principle of the single particle impact test facility [a]

A compressed air connection line on the left and the acceleration tube (length 300 mm, diameter 7 mm) on the right are aligned horizontally, so that the pellet can be accelerated by compressed air (up to 3 bar). When exiting the acceleration tube, the pellet passes the impact cage; top and front side consisting of transparent Perspex plates for optical access. The target plate, which is hit by the pellet, can be adjusted in angles between 20 and 90° by rotation of the axis. The target material can be changed by mounting suitable target plates. A stereoscopic high-speed camera set (2x Mikrotron Cube 7, lens: F1.4 f50 mm) allows for tracking and optical analysis of each individual particle and its impact on the target plate. For avoiding optical distortions, the cameras are mounted at a suitable distance from the impact cage (approx. 1000 mm).

Two exemplary sequences in figures 4.2 and 4.3 show the particle impact and the resulting breakage into two or four fragments from the perspective of the slave and the master camera, respectively.

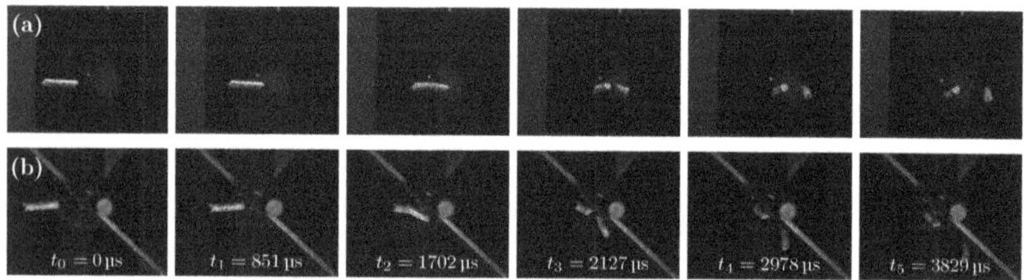

Figure 4.2: Sequence of single particle-wall impact with two resulting fragments: (a) slave camera, (b) master camera ($v_p = 20\,\mathrm{m\,s^{-1}}$, $l_p = 23\,\mathrm{mm}$, $\varphi_i = 45°$)

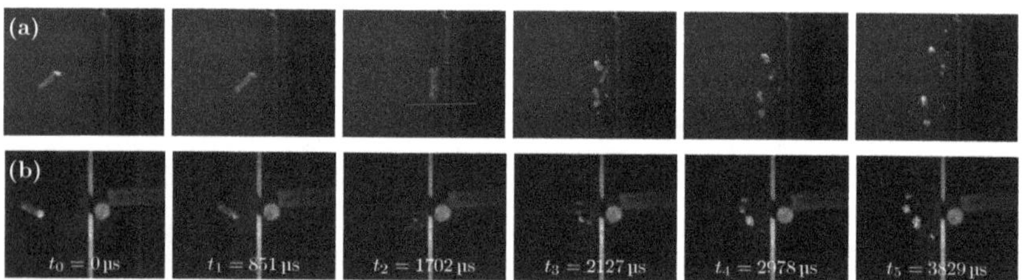

Figure 4.3: Sequence of single particle-wall impact with four resulting fragments: (a) slave camera, (b) master camera ($v_p = 20\,\mathrm{m\,s^{-1}}$, $l_p = 22\,\mathrm{mm}$, $\varphi_i = 90°$)

For simplification, the particle's angular velocity and orientation during impact are neglected in the current contribution, but their influence on the pellets' degradation behaviour is statistically included in the empirical correlations developed. To determine the breakage probabilities (BP) and fragment size distributions for each individual condition, between 10 and 70 impacts were performed, depending on the ambiguousness of the degradation behaviour. The ranges of the test series performed, which the resulting model parameters are based on (listed in tables 4.2 and 4.3), lie between 5 and 40 $\mathrm{m\,s^{-1}}$, 20 to 90° and 5 to 30 mm for impact velocity, collision angle and particle length, respectively.

Stereoscopic particle tracking Each impact is acquired by the stereoscopically arranged high-speed camera set with a frame rate of 2350 fps and a shutter of 35 µs. The stereoscopic tracking enables recording of the spatial trajectories of each particle in three-dimensional space. Digital image processing and analysis is performed by several MATLAB®-scripts developed for this application.
Figure 4.4 provides an example of the resulting particle trajectories. The particle velocity v_x is determined from the change of the particles' position and the frame rate.

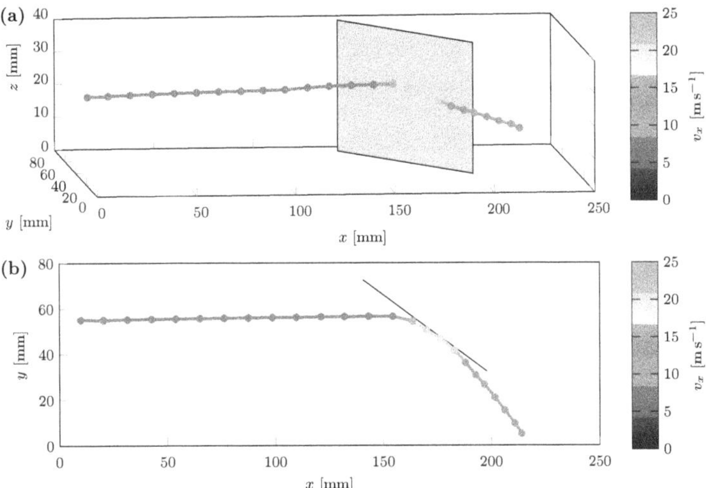

Figure 4.4: Experimentally estimated 3D-trajectories for a single particle impact (target angle $\varphi = 45°$): 3D-view (a) and topview (b)

Four fundamental processing steps are required for tracking particles and detecting their orientation, velocity and position in each frame. The synchronously acquired images are first equalised (*undistortion*) to compensate the possible angular offset and alignment of the two cameras. To adjust the orientation of both cameras digitally (*rectification*), the respective image data are projected into a shared virtual image plane. This allows analytical image analysis for particle detection and tracking. Subsequently, distinctive features are detected in both images (*correspondence*), in this case obtained by tracking. The corresponding 3D-coordinates of each particle position are finally determined by triangulation (*reprojection*). A comprehensive description of the procedure can be found in [121, 179, 230].
Repetitions of the test series show a deviation of 5 % for the detected particle lengths compared to manual measurements and a variation of 5 % for the recorded velocities, which is satisfying in both cases.

In all other empirical studies presented in the literature (section 2.3.4), the fragments are collected and sieved after each test series for mass-weighted determination of the re-

sulting size distribution. However, due to their non-spherical shape, the "true" size of pellets or (in this case) fragments is difficult to quantify by sieving. The particle instability (due to cracks and inhomogeneous structure, investigated by Ochkin-König et al. [161]) would also cause additional size reduction by sieving. Therefore, the implemented digital particle detection and separation routine is employed for counting the number of fragments immediately after impact. A distinction is made between one (i.e. no breakage) up to four (could be more than four) resulting fragments. The detection of a larger number of fragments than four is not sufficiently accurate due to visual particle overlap. By detecting the fragments immediately after impact, any further contact of the fragments with surrounding walls is no longer problematic regarding further comminution.
Table 4.1 provides an exemplary data layout which enables calculation of the percentage of the broken particles and the size distribution of the fragments. Based on the number-related fragment detection, the shares of the appropriate length ratios ($^{l_{p,1}}/_{l_{p,0}}$ - see section 4.3) and thus the resulting mass fractions are determined. Note that for determination of the fragment size distribution the unbroken particles are not taken into account (under size).

Table 4.1: Exemplified experimental data layout after impact ($l_p = 15\,\mathrm{mm}$, $v_p = 35\,\mathrm{m\,s^{-1}}$, $\varphi = 90°$)

No. of fragments [−]	$^{l_{p,1}}/_{l_{p,0}}$ [−]	Detected particles [−]	Weight [g]	Weight fraction [−]	Under size [−]
1	1	0	0	0	1
2	0.5	180	40.2	0.4737	0.5263
3	0.33	180	26.8	0.3158	0.2105
4	0.25	120	13.4	0.1579	0.0526
5	0	50	4.5	0.0526	0
Sum			84.85		

4.2 Materials

In the current thesis, one single pellet type is used. This type was chosen, since the comparatively low mechanical durability (DU) leads to well-measurable comminution effects. The essential material properties like durability, moisture content MC, density ρ_p and the bulk's average particle length $\bar{l}_p$ are listed in figure 4.5. The pellets are labelled as EN*plus*-A1 ($DU > 98.0$, cf. section 2.1).

DU: $(98.50 \pm 0.13)\,\%$

MC: $(6.36 \pm 0.14)\,\mathrm{wt.\text{-}\%}$

ρ_p: $(1053 \pm 30)\,\mathrm{kg\,m^{-3}}$

$\bar{l}_p$: $(12.70 \pm 0.24)\,\mathrm{mm}$

Figure 4.5: Pellets used in the experiments

Note that steel was applied as target material, since the bends installed in the subsequent pneumatic conveying experiments also consist of this material. Therefore, neither the influence of the pellet material nor the target material will be investigated specifically in the current contribution. Information on the influence of these effects on comminution can be found in Jägers et al. [a].

4.3 Development of empirical model

An empirical model for statistical description of characteristics of pellet breakage by impact is developed, based on two essential material functions *selection function* and *breakage function*, referring to breakage probability and the resulting fragment size distribution, respectively (cf. section 2.3.4). The (mostly material-specific) model parameters are determined by model adaption based on the data obtained. The accuracy of the fitting is assessed by the coefficient of determination (R^2), a benchmark how accurately the least squares fitting curve represents the experimental data (values close to 1 correspond to high accuracy). Simultaneously, the influence of impact velocity, collision angle and original particle length on the degradation characteristics are analysed.

4.3.1 Selection function

The selection function (SF) denoted with the letter P further on, describes the probability of particle breakage by impact as a function of impact velocity, collision angle and particle length. Note that in principle, the momentum or kinetic impact energy plays a decisive role in particle degradation behaviour by impact. Since pellets show a constant density with little deviation (cf. figure 4.5), determined for this pellet type and the same batch by Ochkin-König et al. [161] using helium pygnometry, the parameter kinetic impact energy can be reduced to a function of particle length and impact velocity.
Several empirical models exist for describing P. An extensive selection, including e.g. logistic- [102], Weibull- [40, 187] or lognormal-based models [170], was compared by Rozenblat et al. [183]. Since all models are approximately consistent with experimental data, mathematical simplicity is a main reason for particular selection. Since the logistic model parameters have a clear statistical meaning, the model according to Kalman et al. [102] was chosen as selection function and reads as

$$P = 1 - \frac{1}{1 + \left(\frac{v_p}{v_{50}}\right)^{D_i}}, \qquad [-] \qquad (4.1)$$

where P, v_p and v_{50} denote the breakage probability, the impact velocity and the median impact velocity, respectively. The logistic model parameter D_i defines the variance of the data. With increasing D_i, the distribution becomes narrower and converges to a step function when D_i reaches infinity. v_{50} represents the velocity at which half of the particles tend to break by impact. The breakage probabilities P are provided in figure 4.6 (a) for a constant collision angle of 90° and varying pellet lengths as a function of impact velocity. Clearly, the longer the particles are, the higher is P.

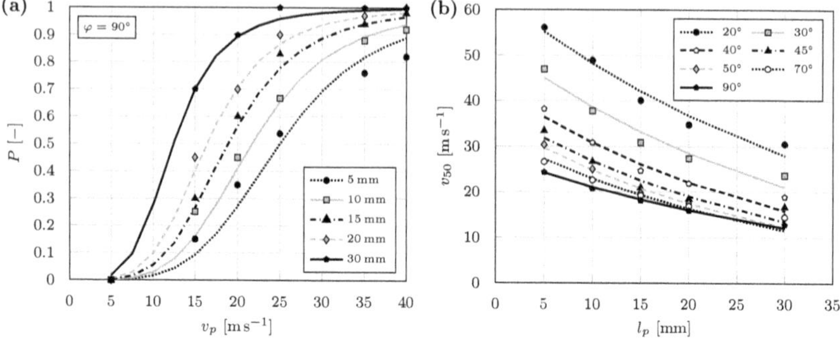

Figure 4.6: Fitting of the experimental data and influence of impact velocity and particle length (a) and dependence of median impact velocity on collision angle (b) [a]

When adjusting the logistic model parameters v_{50} and D_i to all experimental data, it was found that the distribution width D_i does not change significantly with the pellet length, but tends to depend on the collision angle; in contrast to spherical particles, where D_i is purely material-dependent (cf. [183]). Therefore, an almost linear correlation between D_i and the collision angle φ was obtained

$$D_i = a_D\varphi + b_D, \qquad [-] \qquad (4.2)$$

including two empirical parameters a_D and b_D, which are both listed in table 4.2.

v_{50} is plotted against the pellet length l_p for all investigated collision angles (figure 4.6 (b)). The fitting of v_{50} was performed using a correlation of the exponential form

$$v_{50} = A_V \exp\left(-\frac{l_p}{B_V}\right), \qquad [\mathrm{m\,s^{-1}}] \qquad (4.3)$$

where l_p, A_V and B_V denote the mean particle length and two empirical, material-dependent parameters, respectively. The values of both parameters are also listed in table 4.2, leading to an accuracy (R^2) of 0.993. Both A_V and B_V in table 4.2 relate to measurements with a normal collision angle of 90°, an approach corresponding to the suggestions in [22, 23, 223].

Table 4.2: List of the SF logistic model parameters [a]

$D_i = a_D\varphi + b_D$		$v_{50} = A_V \exp\left(-\frac{l_p}{B_V}\right)$	
a_D [$°^{-1}$]	b_D [–]	A_V [m s^{-1}]	B_V [m]
0.0386	0.9817	27.9	0.0356

The collision angle is another significant factor influencing the mean impact velocity (cf. figure 4.6 (b)). Portnikov et al. [171] suggested an exponential model which considers this dependency by determining the ratio of the mean impact velocity v_{50} and v_0 (median impact velocity at 90° collision angle) as follows:

$$\frac{v_{50}}{v_0} = C_V \exp\left(-\frac{\varphi}{\varphi_0}\right) + 1, \qquad [-] \qquad (4.4)$$

where C_V and φ_0 denote constant material and size independent empirical parameters with values of 4.4 and 18.1°, respectively. Figure 4.7 (a) justifies the choice of this correlation since the adjusted data are within a deviation interval of ±20 % for the collision angles between 20 and 90°.

In summary, in accordance with Portnikov et al. [171], the mean impact velocity including equation (4.4) is correlated as follows:

$$v_{50} = A_V \exp\left(-\frac{l_p}{B_V}\right)\left[4.4 \exp\left(-\frac{\varphi}{18.1}\right) + 1\right] \qquad [\mathrm{m\,s^{-1}}] \qquad (4.5)$$

The adaptation of the experimentally determined breakage probabilities to equations (4.1) to (4.3) and (4.5) led to $R^2 = 0.989$. Figure 4.7 (b) provides a comparison of fit and experimental data as a function of impact velocity and different collision angles.

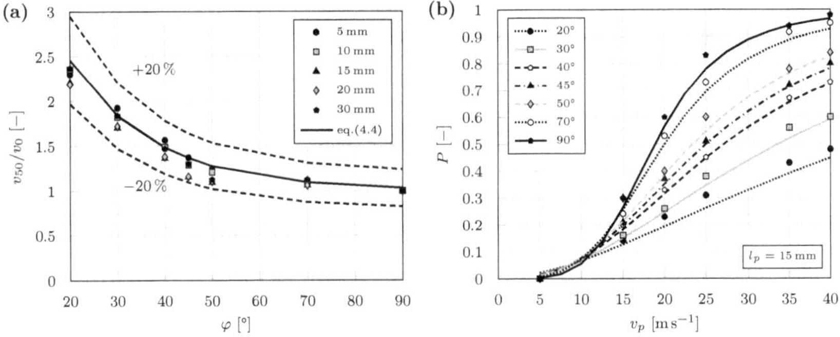

Figure 4.7: Parameter correlation in relation to collision angle (a) and resulting fits of breakage probability for varying collision angles (b) [a]

As indicated in figure 4.6 (b), the breakage probability is clearly reduced at smaller collision angles, even for higher impact velocities.

Figure 4.8 finally summarises the model developments. The experimentally determined breakage probabilities are plotted in figure 4.8 (a) versus the normalised impact velocity in power of D_i in accordance with equation (4.5) for all investigated collision angles and particle lengths. The solid line denotes a perfect fit of equations (4.1) and (4.5) to experimental data. The chart confirms a model accuracy of 20 %, even for cylindrical wood pellets.

Both experimentally determined and calculated breakage probabilities are plotted in figure 4.8 (b). A good overall accuracy of ±20 % can be stated. In both diagrams discrepancies mainly occur for smaller particles. Due to the cylindrical shape, extraordinary impact scenarios, where the pellets are compressed instead of breaking, can occur at critical impact velocities especially at collision angles of 90°. These scenarios cannot be modelled in the logistic model applied (equation (4.1)).

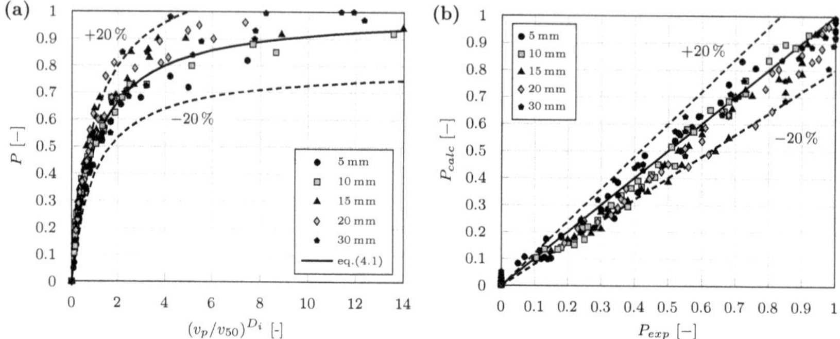

Figure 4.8: Validation of the logistic function model (a) and comparison of the predicted and the experimental breakage probabilities (b) [a]

4.3.2 Breakage function

The breakage function (BF) denoted with the letter B further on, describes the mass-based size distribution of the fragments. Similar to SF, it depends on impact velocity, collision angle and size (or length) of the initial particle. For predicting the size distribution of resulting fragments, several BF models [9, 40, 82, 198, 211, 226] were compared by Rozenblat et al. [183]. Based on the coefficients of determination (R^2), the model according to Tavares et al. [211] was selected for describing the fragments size distribution. It was modified with regard to the current particle size determination method via digital image analysis. Here, the ratio between the length of the initial mother particle $l_{p,0}$ to the detected fragment length $l_{p,1}$ replaces the ratio between the initial parent sieve size interval and the representative size of fragments in the respective class (cf. [211]). Note that for breakage function determination, the unbroken particles were not considered (i.e. under size). The model reads as:

$$B = 1 - (1 - t_{10})^{\left(\frac{9}{l_{p,0}/l_{p,1} - 1}\right)^{\alpha}}, \quad \text{for} \quad l_{p,1} < l_{p,0} \qquad [-] \qquad (4.6)$$

where B, t_{10} and α denote the cumulative mass distribution and two model parameters, respectively.

Figure 4.9 provides the cumulated size distribution (under size) B of the resulting fragments as a function of $l_{p,1}/l_{p,0}$ for a collision angle of 90°. The mass-based cumulative distribution was derived from optical size determination using the material density (see figure 4.5). Figure 4.9 reveals that the larger the original particle size or the higher the impact velocity, the smaller the size and larger the number of fragments. According to Rozenblat et al. [183] and Antonyuk et al. [8], this behaviour is consistent to other materials such as salt, potash or zeolite, despite of the very inhomogeneous material structure and particle shape of pellets. The data were adjusted according to equation (4.6) for determining the model parameters t_{10} and α, respectively. Thereby, minimum accuracies (R^2) of 0.987 (figure 4.9 (a)) and 0.971 (figure 4.9 (b)) were achieved, respectively.

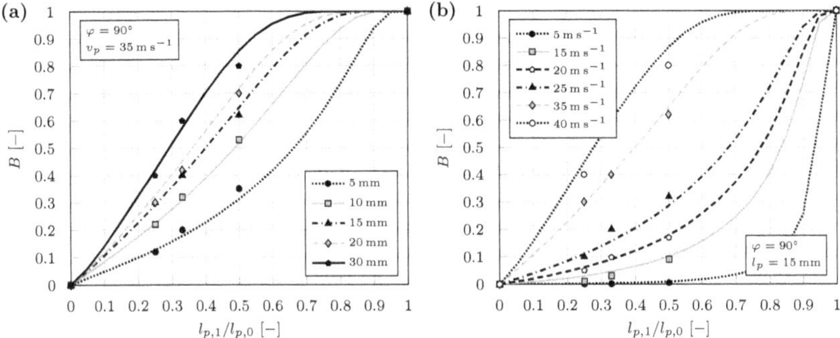

Figure 4.9: Fitting of the experimental data: Dependence of fragment size distribution on original particle length (a) and impact velocity (b) [a]

The model parameter t_{10} in the Tavares model [211] denotes the fraction of fragments that make up 10 % of the initial particle size. Rozenblat et al. [183] recommended the following correlation approach for t_{10}

$$t_{10} = 1 - \exp\left(-a_t v_p^{b_t} l_p\right), \qquad [-] \qquad (4.7)$$

where, in addition to the original particle size l_p and the impact velocity v_p, a_t and b_t denote two material and particle size-specific model parameters.

Figure 4.10 (a) depicts t_{10} as a function of the impact velocity for all pellet lengths investigated. The data show that increasing impact velocities cause lager amounts of small fragments (i.e. fines). During evaluation, it was found that a_t and b_t do not only depend on the material properties as in case of spherical particles like e.g. salt, potash, zeolite, but in case of cylindrical wood pellets also on the original particle length. Hence, a_t and b_t were determined using linear or polynomial length-based correlations (equations (4.8) and (4.9)) with material-specific parameters a_1, a_2, b_1, b_2 and b_3 ($R^2 = 0.985$).

$$a_t = a_1 l_p + a_2, \qquad [\mathrm{s\,m^{-2}}] \qquad (4.8)$$

$$b_t = b_1 l_p^2 + b_2 l_p + b_3. \qquad [-] \qquad (4.9)$$

The values of the parameters obtained by fitting a_t and b_t are listed in table 4.3.

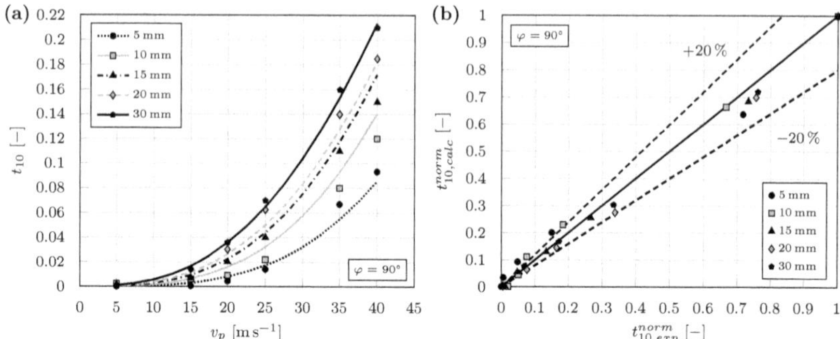

Figure 4.10: Fitting and dependence of model parameter t_{10} on impact velocity and original particle length (a) and corresponding parameter validation (b) [a]

The second parameter α of the BF model was determined by fitting the fragment size distribution (see figure 4.9) and is given in figure 4.11 (a) for different mother particle lengths as a function of impact velocity. In case of cylindrical pellets, a polynomial correlation (2nd degree) between α and the impact velocity was obtained:

$$\alpha = c_\alpha v_p^2 + d_\alpha v_p + e_\alpha \qquad [-] \qquad (4.10)$$

For wood pellets, the model parameters (otherwise assumed in literature as material-specific) also depend on the pellet length (due to material inhomogeneity). A linear dependence on the original particle size (with a minimum R^2-value of 0.969) was found for the individual model parameters c_α, d_α and e_α, respectively:

$$c_\alpha = c_1 l_p + c_2, \qquad [\mathrm{s^2\,m^{-2}}] \qquad (4.11)$$

$$d_\alpha = d_1 l_p + d_2, \qquad [\mathrm{s\,m^{-1}}] \qquad (4.12)$$

$$e_\alpha = e_1 l_p + e_2. \qquad [-] \qquad (4.13)$$

Figure 4.11 (a) provides the fitting curves for α. Although the underlying models for α are different for wood pellets and for other granular material such as salt, potash or zeolite [183] (polynomial correlation instead of power law relationship), the same trend can be observed; α increases with particle size. Table 4.3 gives an overview of all parameter values obtained.

Table 4.3: List of the BF model parameters [a]

$t_{10} = 1 - \exp\left(-a_t v_p^{b_t} l_p\right)$						
$a_t = a_1 l_p + a_2$		$b_t = b_1 l_p^2 + b_2 l_p + b_3$				
a_1 [s m^{-3}]	a_2 [s m^{-2}]	b_1 [m^{-2}]	b_2 [m^{-1}]	b_3 [−]		
0.0133	-4×10^{-5}	816.9	−59.61	3.7669		
$\alpha = c_\alpha v_p^2 + d_\alpha v_p + e_\alpha$						
$c_\alpha = c_1 l_p + c_2$		$d_\alpha = d_1 l_p + d_2$		$e_\alpha = e_1 l_p + e_2$		
c_1 [s^2 m^{-3}]	c_2 [s^2 m^{-2}]	d_1 [s m^{-2}]	d_2 [s m^{-1}]	e_1 [m^{-1}]	e_2 [−]	
0.0781	7×10^{-5}	−4.8257	0.0049	83.398	0.6154	

Finally, the normalised experimentally determined values are compared with the correlations developed (equations (4.10) to (4.13)). The results are provided in figure 4.11 (b). Note that due to the polynomial dependence of α on the impact velocity, the normalised values (by division by the α-value of the maximum impact velocity) can exceed the value of 1. Again, the values calculated remain within an acceptable deviation sector of ±20 %.

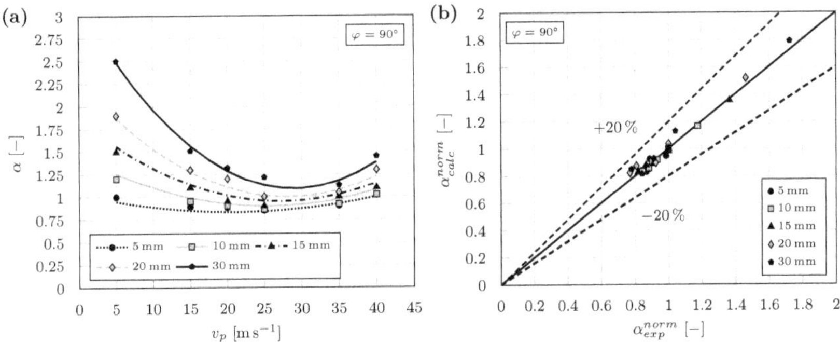

Figure 4.11: Fitting and dependence of model parameter α on impact velocity and original particle length (a) and and corresponding parameter validation (b) [a]

For the final verification of the correlations obtained, experimental results of B were compared to the predicted ones. The results, provided by figure 4.12 and including all impact velocities investigated, indicate a good agreement within a tolerance range of ±25 %. Note that the larger deviations only occur in the range of B from 0 to 0.2 mainly for larger mother particle lengths. Smaller values of B represent smaller fragment sizes (cf. figure 4.9), and hence a higher fragment quantity for larger mother particles. Thus, a larger deviation between the calculated and experimentally determined data could therefore result from the fact that not all fragments were detected by the stereoscopic high-speed cameras.

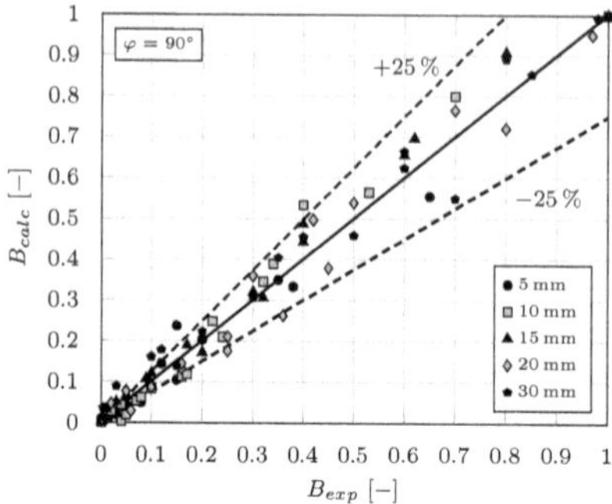

Figure 4.12: Comparison of predicted and experimental fragment size distributions [a]

Note that the BF was previously determined as a function of impact velocity and original particle length, but only for impacts in normal direction (i.e. 90°). The effect of the collision angle was neglected so far. The procedure to account for the dependence of BF on collision angle is similar to the one previously applied for predicting SF. Principally, the impact velocity is reduced by equation (4.14) (similar to equation (4.4) for breakage probability prediction). Due to its validity for SF, this has only been assumed by Portnikov et al. [172] for predicting BF, but was not analysed in detail so far.

$$\frac{v_p}{v_{90}} = C_V \exp\left(-\frac{\varphi}{\varphi_0}\right) + 1 \qquad [-] \qquad (4.14)$$

Similarly to SF, C_V and φ_0 amount to 4.4 and 18.1°. v_{90} represents the horizontal impact velocity at 90° and v_p denotes the resulting horizontal impact velocity affected by collision angle.

The effect of the collision angle on B is shown in figure 4.13.

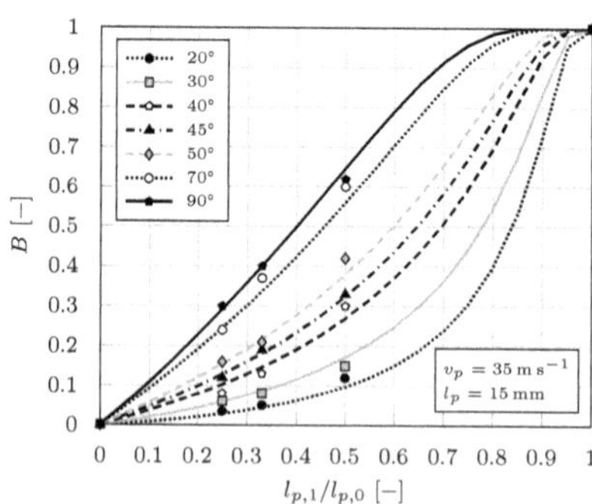

Figure 4.13: Fitting of the experimental data: Influence of collision angle on fragment size distribution [a]

The plots correspond to the predicted values using equation (4.14), merged with the previously validated model (equations (4.6) to (4.13)). The predicted values agree with the experimental data with an accuracy of 0.991 (R^2). From physical point of view, flatter collision angles result in a reduced probability of smaller fragments being produced, which is in line with the assumption that a flatter collision angle reduces the relevant (normal) component of the impact velocity.

Exemplary for a constant initial pellet length for all investigated relevant normal velocities between 5 and 40 m s^{-1}, the dependence of the angle-velocity correlation (equation (4.14)) on both individual model parameters t_{10} and α is further examined (figure 4.14). Both calculated normalised values of t_{10} (a) and α (b) are in reasonable agreement with the experimental values and lie within acceptable intervals of ±20 %.

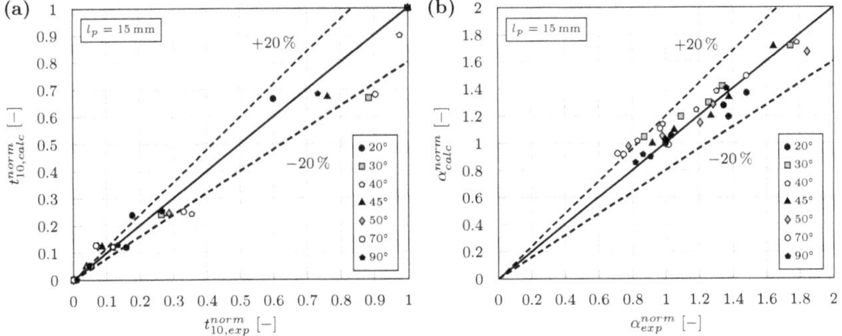

Figure 4.14: BF-model parameter validation for varying collision angles: t_{10} (a), α (b) [a]

For verifying the accuracy of the resulting model for predicting the cumulative size distribution B considering the influence of collision angle, the predicted values of B were plotted against the experimental results, exemplary for one pellet length (figure 4.15). The results indicate a satisfying overall accuracy with a deviation interval of ±25 %.

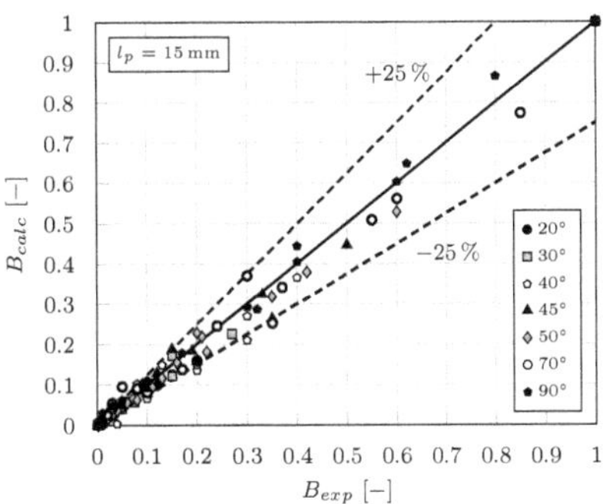

Figure 4.15: Validation of the predicted fragment size probabilities concerning varying collision angles [a]

4.4 Numerical investigation

The developed empirical functions for predicting particle breakage probability (P) and resulting fragment size distribution (B) are transferred into the in-house DEM code for subsequent investigations into size reduction by pneumatic conveying. Prior final validation, the implementation of the model including the fragment size selection and the discretised particle degradation are described.

4.4.1 Implementation in DEM

Figure 4.16 provides the principle of the proposed numerical degradation algorithm. After initialisation and particle generation, fluid flow (in case of coupled DEM-CFD scenarios) and particle motion are determined using the DEM-CFD coupling approach. In case of particle-wall collision, the relevant collision parameters are transferred to the degradation algorithm. The decision on particle breakage is based on the selection function (cf. section 4.3.1). In case of particle breakage, number and size of the resulting fragments are determined by the breakage function (cf. section 4.3.2). Afterwards, the fragments are generated and their properties are transferred back to the DEM-CFD algorithm. There, the fluid properties and particle positions are updated. When the final calculation time is reached, the simulation is terminated.

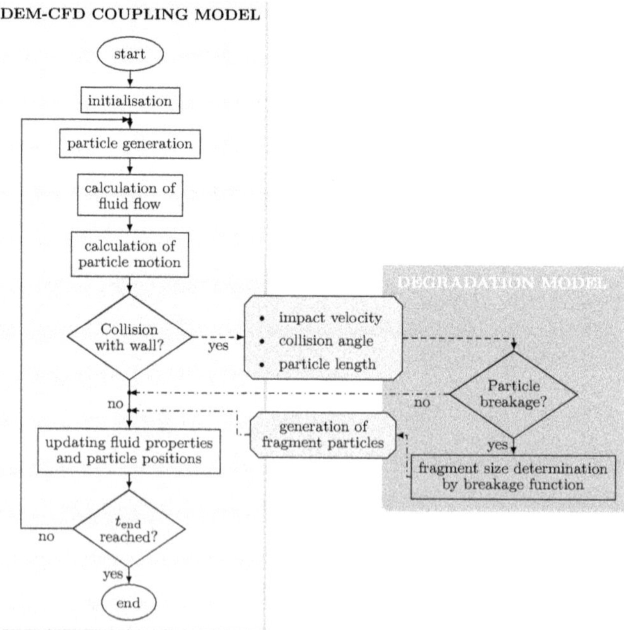

Figure 4.16: Schematic of particle degradation algorithm within DEM-CFD coupling approach

Decision on particle breakage In case of a particle-wall contact, the corresponding breakage probability P (value between 0 and 1) is determined using the empirical selection function as a function of the input conditions (impact velocity, collision angle and initial particle length). A random number between 0 and 1 is generated using a pseudo-random integral number generator [142], which returns a sequence of appearing unrelated numbers at each call. If the generated random number is less than the calculated probability, particle breakage is assumed. Otherwise the particle remains unbroken.

Selection of fragment sizes The breakage function delivers a mass-based size distribution (equations (4.6), (4.7) and (4.10)). Using a mass-based size distribution in the numerical algorithm would lead to a significant error, since the breakage algorithm is adapted for individual particles and the fragment size determination procedure requires a random selection of discrete particles. Thus, the mass-based size distribution needs to be converted into a number-based distribution. Since the cumulative mass-based size distribution (B) is a continuous function, the conversion can be performed as follows [103]:

$$N = \int_{l_{min}}^{l} \frac{b/l}{\int_{l_{min}}^{l_{max}} b/l \cdot \partial l} \partial l, \qquad [-] \qquad (4.15)$$

where N, l, l_{max} and l_{min} refer to the cumulative numbered fragment ratio, the fragment length, the maximum particle length (size of initial particle) and the minimum fragment size, respectively. b denotes the mass-based fragment frequency ratio:

$$b = \frac{\partial B_{i,j}}{\partial l} \Delta l \qquad [-] \qquad (4.16)$$

Since the analytical determination of N could be complicated depending on the derivation complexity of B, it is preferable to solve equation (4.15) by discretisation. B is therefore divided into discrete length intervals for each possible mother particle length (between 6 and 35 mm in 1 mm intervals, see section 5.2.1). The cumulative numbered fragment ratio N is calculated according to discrete expression of equation (4.15) [172].

$$N_i = \frac{\sum_{k=1}^{i} b_k/l_k}{\sum_{k=1}^{j} b_k/l_k} \qquad [-] \qquad (4.17)$$

Index i denotes the i-th size interval and j the respective largest possible size depending on the mother particle. l_k refers to the length of the specific size fraction k and b_k represents the differential of B:

$$b_k = B_{k+1} - B_k \qquad [-] \qquad (4.18)$$

For saving computational efforts during simulation, the cumulative numbered fragment ratio was estimated using equations (4.17) and (4.18) for all possible collision scenarios (mother particle lengths, collision velocities and angles) in advance during initialisation process. The appropriate values were tabulated for subsequent repeating access in intervals of 1 mm, 5 m s^{-1} and 5°, respectively.

After estimating the cumulative numbered fragment ratio of a specific mother particle at given impact velocity and collision angle using equations (4.17) and (4.18), the lengths of each individual fragment needs to be predicted. Similar to breakage probability, this was performed by generating a random number between 0 and 1 for N and interpolating between the pre-calculated discrete values. The random selection is repeated until the total length of the selected fragments reached the mother particle length. In case the total length of the fragments exceeds the initial length, the length of the last fragment is adjusted to the missing length. This procedure is sketched exemplary in figure 4.17.

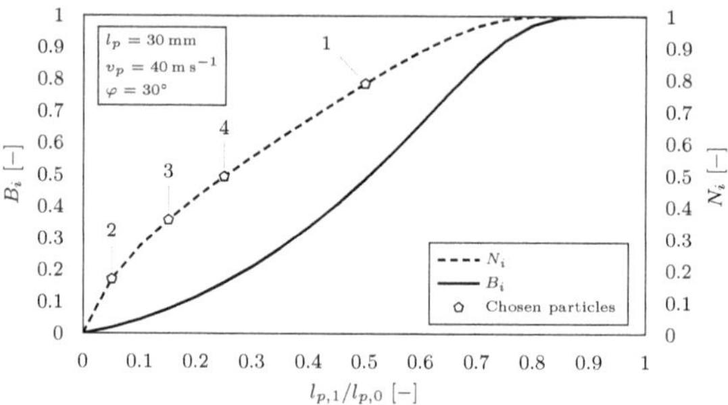

Figure 4.17: Exemplified fragment size selection for a broken pellet

The curve of the mass fragment ratio is predicted according to equations (4.6), (4.7) and (4.10) and the numbered fragment ratio is plotted using equations (4.17) and (4.18). Note that the numbers refer to the order of the randomly selected fragments. Again, the graphical representation underlines the importance of predicting the fragment sizes according to the number-based instead of the mass-based fragment ratio.

Particle degradation modelling After selecting the number and size of fragments, these are generated as spherocylindrical or spherical particles or (virtually) acquired as fines. As long as the length of the fragments is larger than the particle diameter d_p, the fragments are represented as spherocylinders. If the fragment length lies between $d_p/2$ and d_p, the fragments are approximated as spheres with a diameter of $d_p/2$ (i.e. 3 mm). This procedure is similar to the experimental length determination method, where for all optically measurable particles with a length shorter than the original diameter a length of half of the diameter (i.e. 3 mm) is assumed (see section 5.1.1). All fragments with a length less than half of the pellet diameter are stored but not further treated as DEM particles. The procedure of generating the fragments and calculating the resulting fraction of fines is provided in figure 4.18 on the basis of the exemplary selected fragments in figure 4.17. In figure 4.18 (b), the fragment selected first (1) with a length of 15 mm is approximated as spherocylinder. Contrary, the following fragment (2) with a length of 2 mm is too short, neither to be generated as spherocylinder with initial pellet diameter nor as spherical particle (diameter of 3 mm). Instead, it is captured as virtual fines, but not represented as a DEM particle (figure 4.18 (b), green). The third fragment (3) with a total length of 5 mm is too short for spherocylindrical approximation. Thus, it is split into a spherical particle (diameter of 3 mm) and virtual fines (green) with a remaining length of 2 mm. The last fragment (4) has a remaining length of 8 mm. Consequently, the mother particle exemplary shown in (a) results in the fragments provided in (c).

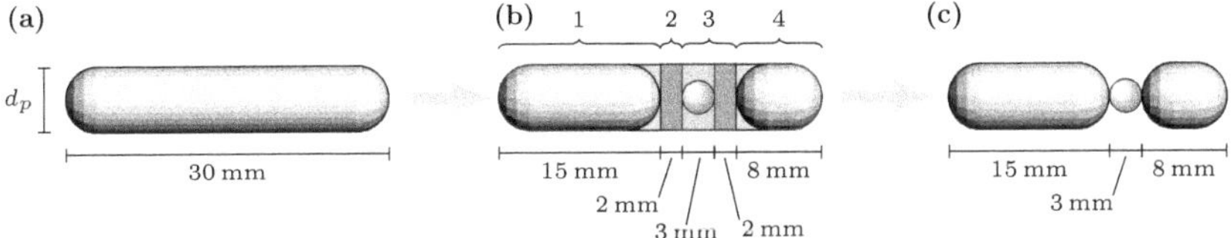

Figure 4.18: Principle of fines determination and generation: initial mother particle (a), selected fragments including virtual fines (b) and resulting generated fragments (c)

Both the volume differences of spherocylindrical fragments (blue) and the volumes of short cylindrical fragments (virtual fines, green) are considered for virtual fines determination. Therefore, the mass of fines produced is determined by the volume difference of the mother particle V_{init} (a) and the resulting fragments V_{frag} (spherocylinders) or V_{sph} (spheres) (cf. (b)). Note that since the spheres represent fragments with an average length of 3 mm, the remaining volume (orange) is not considered for virtual fines determination. Consequently, the volume difference determines the volume and thus the mass of the virtual fine fraction:

$$m_f = \rho_p \cdot V_f = \rho_p \left[V_{init} - \left(\sum V_{frag} + \sum V_{sph} \right) \right] \qquad [\mathrm{kg}] \qquad (4.19)$$

For fragment generation, certain properties (e.g. position, velocity, rotational velocity) are essential for initialisation. Figure 4.19 demonstrates the principle of splitting the mother particle after particle-wall collision and the subsequent generation of the resulting fragments including the initial particle properties.

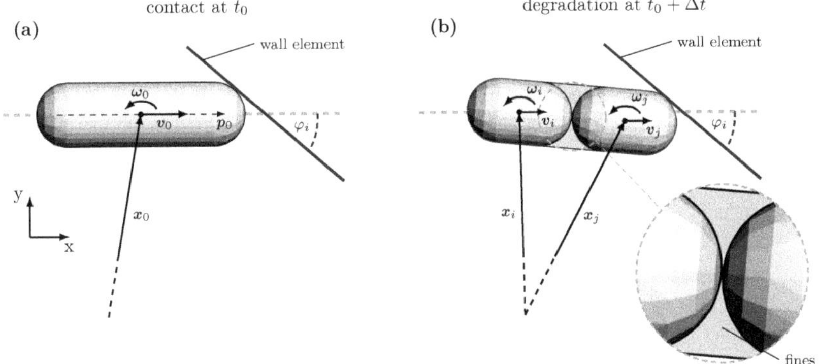

Figure 4.19: Working principle of pellet degradation algorithm: particle wall contact at impact angle φ_i and time t_0 (a) and particle split into fragments i, j and virtual fines at time $t_0 + \Delta t$ (b)

After particle-wall collision at time t_0 (a), the mother particle is in this case divided into two fragments i and j in the following time step $t_0 + \Delta t$ (b). The initial properties of both fragments are based on those the mother particle would have at this time step after collision. This procedure ensures that the kinetic energy loss due to impact is still taken into account. Note that the energy loss due to particle breakage is neglected at this point.

The initial fragment position $\boldsymbol{x}_k$ ($k = i, j \dots n$) and orientation $\underline{q}_k$ are determined in dependence of the mother particles position $\boldsymbol{x}$ and the cylinder axis $\boldsymbol{p}$ at time $t_0 + \Delta t$:

$$\boldsymbol{x}_k = \begin{cases} \boldsymbol{x}_{0+\Delta t} + \frac{1}{2} \frac{\boldsymbol{p}_{0+\Delta t}}{|\boldsymbol{p}_{0+\Delta t}|} \left(l_{p,0} - l_{p,k} \right) & k = i \\ \boldsymbol{x}_{k-1} + \frac{1}{2} \frac{\boldsymbol{p}_{0+\Delta t}}{|\boldsymbol{p}_{0+\Delta t}|} \left(l_{p,k-1} - l_{p,k} \right) & k = j \dots n \end{cases} \qquad [\mathrm{m}] \qquad (4.20)$$

$$\underline{q}_k = \underline{q}_{0+\Delta t} \qquad [-] \qquad (4.21)$$

Here, $l_{p,0}$, $l_{p,k}$ and $\underline{q}_{0+\Delta t}$ denote the lengths of the initial particle and the resulting fragment k and the orientation of the mother particle at the following time step, respectively. The assumed initial fragment velocities $\boldsymbol{v}_k$ are determined as follows:

$$\boldsymbol{v}_k = \boldsymbol{v}_{0+\Delta t} + \boldsymbol{\omega}_{0+\Delta t} \times \left(\boldsymbol{x}_k - \boldsymbol{x}_{0+\Delta t} \right) \qquad [\mathrm{m\,s^{-1}}] \qquad (4.22)$$

They are therefore dependent on the new particle position and the mother particle's position, velocity and rotational velocity $\boldsymbol{\omega}_{0+\Delta t}$ at time of breakage. The fragments' initial rotational velocities are calculated as follows:

$$\boldsymbol{\omega}_k = \frac{\left(\boldsymbol{x}_k - \boldsymbol{x}_{0+\Delta t} \right) \times \left(\boldsymbol{\omega}_{0+\Delta t} \times 2 \left(\boldsymbol{x}_k - \boldsymbol{x}_{0+\Delta t} \right) \right)}{\left| \boldsymbol{x}_k - \boldsymbol{x}_{0+\Delta t} \right|^2} \qquad [\mathrm{rad\,s^{-1}}] \qquad (4.23)$$

The virtual acquisition of the resulting fines ensures the conservation of momentum:

$$m_0 \boldsymbol{v}_{0+\Delta t} = \boldsymbol{v}_{0+\Delta t} \sum_{k=1}^{n} m_k + \cancelto{\text{fines}}{m_f \boldsymbol{v}_{0+\Delta t}} \qquad [\mathrm{N\,s}] \qquad (4.24)$$

By applying equations (4.20) to (4.23), resulting fragments first behave like the mother particle at the time step of degradation $(t_0 + \Delta t)$ before they are moving on according to their own mechanical and physical laws.

4.4.2 Verification

The breakage probabilities and fragment size distributions numerically predicted by the implemented degradation model are subsequently compared to the experimental and calculated results for varying collision scenarios (particle length, collision velocity and angle). Therefore, the experimentally performed single particle impact scenarios are performed correspondingly, each with 200 impacts. The resulting breakage probabilities and fragment size distributions are determined in analogy to the experimental determination procedure.

Figure 4.20 illustrates in (a) the simulated breakage probability distributions as a function of collision velocity compared to the experimental and calculated values. In (b), the values for different particle lengths are evaluated. Note that the values provided contain different collision velocities and angles, respectively.
The numerically achieved data reflect both the calculated and the experimentally determined breakage probabilities with acceptable accuracy. Only for small collision angles (e.g. 30° in (a)) the values diverge from the calculated ones, but agree well with experimental values. Main reason is the dependence of the probability on the collision angle (cf. equation (4.5)) and thus the flat trend at flat collision angles. In figure 4.20 (b), the experimentally determined and calculated breakage probabilities almost lie within a deviation interval of ±20 % and show acceptable agreement with the numerical values.

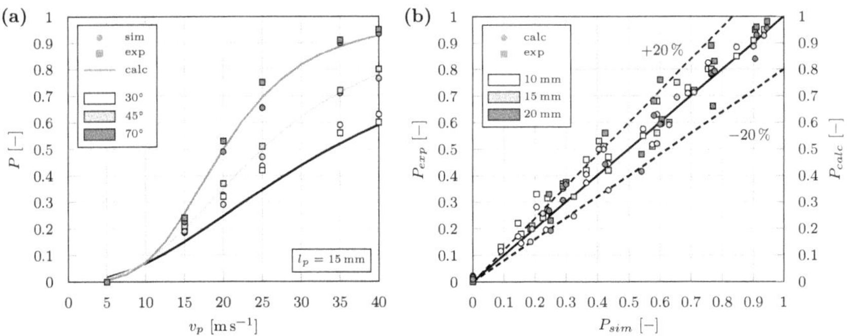

Figure 4.20: Comparison between numerical, experimental and calculated breakage probabilities for different collision angles (a) and particle lengths (b)

Results in figure 4.21 show reasonable agreement of the simulated fragment size distributions with experimentally determined and calculated values for different collision angles, in detail for one scenario of constant impact velocity and particle length as a function of collision angle (a) or more conclusively including further collision velocities (b).

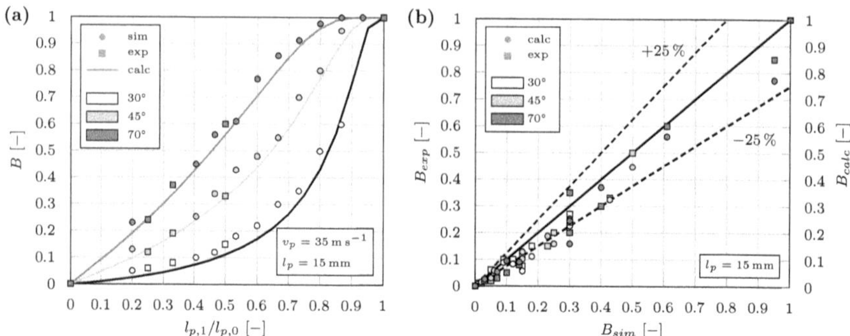

Figure 4.21: Comparison between numerical, experimental and calculated fragment size distributions (a) and probabilities (b) for different collision angles

Compared to the experimentally determined or calculated values, the numerical results widely show acceptable deviation of at most ±25 %. Deviations between the calculated and simulated values are mainly caused by the particle degradation principle applied within the numerical degradation model (e.g. approximation of smaller fragments as cylindrical slices).

5 Experimental and numerical setup for pneumatic conveying analysis

For the experimental analysis of wood pellet degradation during pneumatic conveying and the resulting pressure losses, a test rig was developed allowing for investigation into the influences of several operating conditions. In addition, an optical measuring system was designed which enables the length measurement of larger pellet quantities and thus the quantification of the degradation effects. In addition to the experimental procedure, both systems are described subsequently, together with the numerical setup for theoretical investigation. Finally, the selected numerical models and assumptions are verified by comparison of representative experiments and simulations, e.g. silo discharge and settling behaviour of a single particle.

5.1 Experimental setup

5.1.1 Optical length measuring system

The developed measuring device and the corresponding digital image analysis routine for pellet length determination is described in detail in the contributions of Jägers et al. [g, j]. Thus, only the essential processing and analysis steps are explained subsequently.

Design and operating principle The 3D-model and operating principle of the measuring device are sketched in figure 5.1. The pellets, initially stored in the silo, are successively dosed on a LED-panel (squared, 600 mm side length) through a feeding system. This consists of a cone, vertically moved by a clocked pneumatic cylinder (figure 5.1, red circle). Dimensions have been selected to leave a gap between cone and adjacent pipe section in the extended (i.e. closed) state ensuring a gentle stop of the pellets without damages. While falling through the subsequent pipe section, small particles and fines contained in the sample are vacuumed off through a 3.15 mm-mesh (figure 5.1, orange circle). The fines-suctioning air stream is regulated by a bypass. Afterwards, the fines are collected by a cyclone and finally measured gravimetrically. The pellets (100 to 200 per step) slip down a distribution slide onto the LED-panel, surrounded by a damper and side walls. Note that due to their cylindrical shape, the pellets have a tendency to roll.

In combination with the distribution slide installed, this prevents an overlapping of the pellets on the panel.

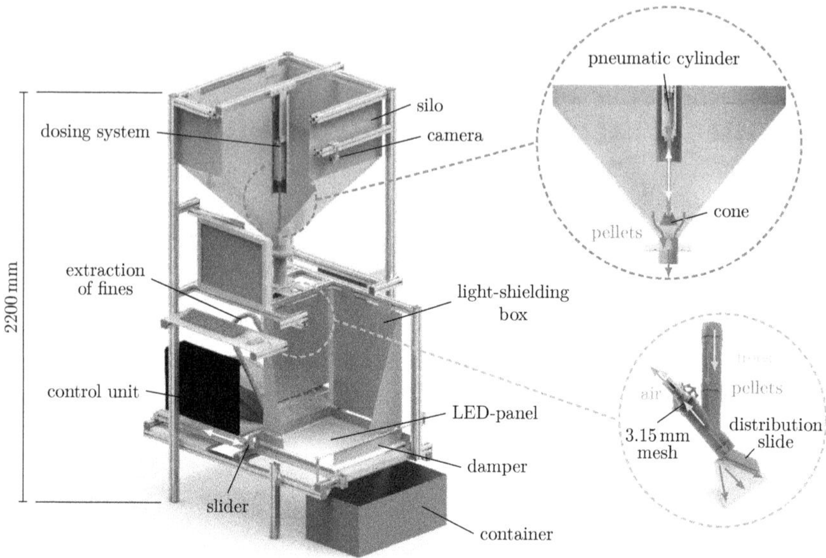

Figure 5.1: Design and working principle of the optical length measuring system [g]

The LED-panel is covered by a light-shielding box in order to avoid falsifying ambient light falling onto the panel and the pellets, which would distort the detected contours of the particles. An area scan camera (Basler acA4600-10uc, lens: F1.8 f12mm) captures an image of the contours of the backlit pellets lying on the panel. Since the pellets fall randomly onto the LED-panel, their position and orientation vary from image to image. Depending on the optical arrangement (perspective of the camera lens in relation to each pellet), optical distortion can occur during 2D-image acquisition. This can lead to inaccuracies, especially in the current case of length determination. For instance, mistakenly detected front surfaces of the cylindrical pellets can result in too large particle length due to parallactic deviation. This is prevented by the comparably large distance between the camera lens and the LED-panel (1500 mm).

Finally, a slider automatically removes the pellets from the panel and the process of acquiring images of pellet contours restarts. The captured images are analysed simultaneously by the image processing routines described in the following section.

Particle detection and length measurement Image segmentation is a crucial step in object contour analysis. Segmentation can be achieved by detection and correct linking of concave points (e.g. Zafari et al. [248], nanoparticles on microscopic images, Yao et al. [244], rice quality control in food industry) or by application of various digital filters and transformations such as Canny edge algorithm combined with the Hough transformation (e.g. Meng et al. [146], analysis of images from transmission electron microscopy) or the watershed transformation. The watershed transformation [149] is used for example by Zheng et al. [252] for the segmentation of contacting soil particles or by Biswas et al. [15] in combination with a Sobel filter for blood cell detection. For pellet contour detection and separation watershed segmentation is applied, since it is known to provide reliable results even if the image contrast is poor [106, 108].

As a first step of each measurement, the optical system is calibrated by determining a conversion factor for each pixel of the ROI (region of interest) in spatial dependence using a mask (consisting of a thin transparent plate with 2025 circular dots of known size). This is essential for the subsequent length determination of the pellets, since the respective pixels are assigned to the corresponding object. In this way, a pixel-precise conversion of the detected area into square millimetres is achieved.

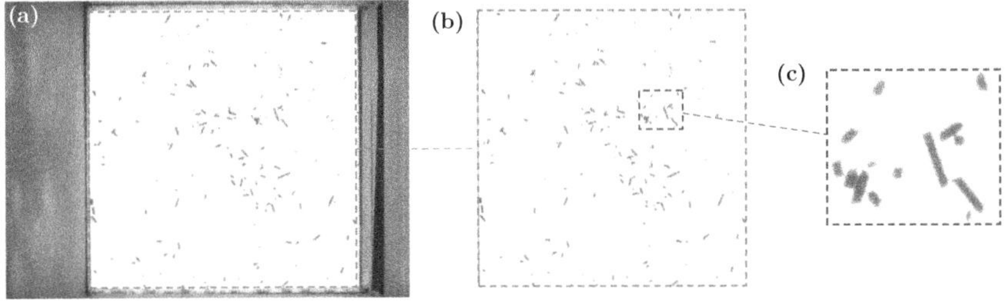

Figure 5.2: Original RGB-image of the LED-panel with pellets (a), cropped ROI-image (b) and illustrative subset of ROI (c) [g]

After calibration, the region of interest (ROI) is detected. Figure 5.2 (a) depicts the initial camera image showing the pellets lying on the LED-panel and parts of the light-shielding box. The cropped segment for later image processing is provided in (b). An illustrative subset showing the randomly distributed pellets in more detail is indicated by (c). Note that in the following, this subset is used to explain the implemented image processing and analysis steps.

The contour images are analysed in several steps using a MATLAB®-script to detect and separate the pellets from the background and for final length determination. The procedure is illustrated by a series of images in figure 5.3. To obtain sharp contours, contrast and color scheme are globally adjusted in (b) and (c) before the image is modified by a first watershed transformation step [149], revealing the positions and orientations of the detectable particles.

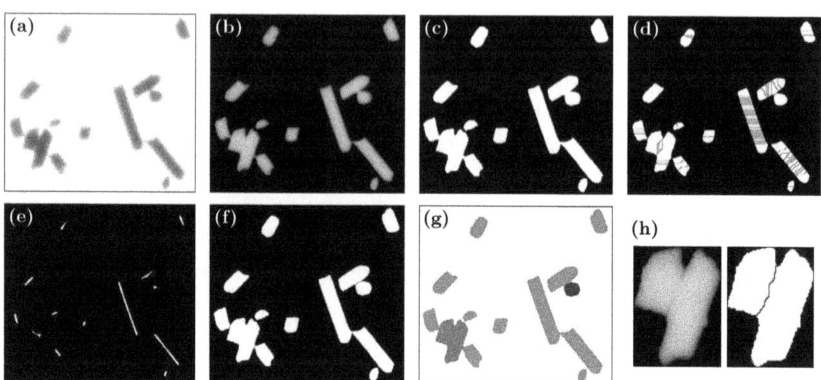

Figure 5.3: Steps of image processing: raw image (a), substraction with empty panel (b), BW-image (c), watershed transformation (d) and (e), seperated particles (f), resulting RGB-image (g), result of intensified processing (h) [j]

The result of this transformation is combined with the binary representation (c), which yields in (d). The resulting image again is treated by a watershed transformation with a second set of parameters (e). Combining this with the binary representation (c) yields the image shown in (f). Adding colour information for separated particles results in (g). As pellets with a length shorter than their diameter tend to fall onto their cross-sectional area, they can be identified by their circularity ((g), marked in blue). Pellets, which are identified correctly, are marked in red (g). Some particles that may lay too close or be in touch, are incorrectly identified as one particle ((g), marked in green). This artefact is removed by an additional image processing step as shown in (h). Finally, the pellet length is determined by dividing the detected area by the minor axis length (diameter) of the respective particle.

Reproducibility of length measurements Measurements were carried out with different test particles: samples mixed from test cylinders of massive wood with 6, 18 and 30 mm length and wood pellets, having different length distributions. All test particles or pellets have a diameter of 6 mm. Using these test particles or pellets, the accuracy of the pellet length measuring method was determined through various test series. The parameters examined are:

- Influence of pellet position and orientation on the LED-panel,
- Accuracy of the fines determination,
- Crushing effect of the measuring device,
- Influence of the detected circular objects.

Due to the greater influence, the first two parameters are described in detail below, in accordance to [g]. The latter two parameters are briefly summarised afterwards.

In the following analysis, the arithmetic average length $\bar{l}_p$ of each sample, as defined in equation (5.1), is applied as an indicator for pellet degradation.

$$\bar{l}_p = \frac{1}{n}\sum_{i=1}^{n} l_{p,i} \qquad [\mathrm{m}] \tag{5.1}$$

The effect of pellet position and orientation on the accuracy of the length measurement was examined by manual positioning of the pellets on the panel (manually, to exclude the mechanical impact of the feeding system) and following automatic analysis by the optical system. Contours of the same pellets were acquired three times after manual re-positioning. For comparison, the samples were measured manually (again three times) with a caliper. This process was performed for five different samples of pellets and test cylinders (TC) of various size distributions. The average lengths of the samples determined are provided in figure 5.4. The three resulting average lengths of each of the individual measurements for the first sample (1) are provided separately in both (a) and (b). The three lengths were combined for the remaining samples (2-5) and plotted together with the respective standard deviation.

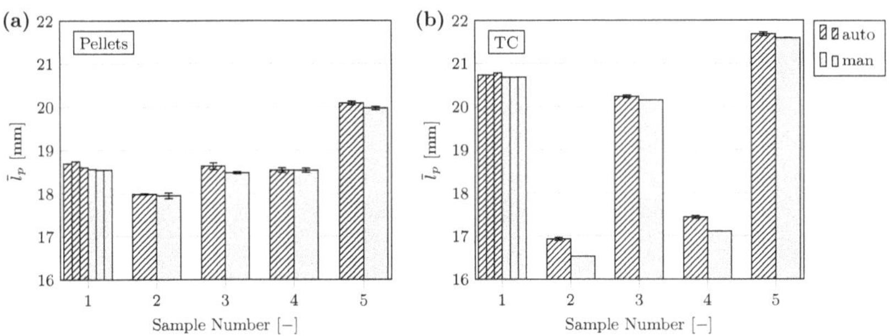

Figure 5.4: Reproducibility of optical length measurement: Dependence on pellet position and orientation and comparison between automatic and manual measurements for pellets (a) and test cylinders (b) [g]

The standard deviations (error bars) are small in all of the five automated measurements, independent whether wood pellets or test cylinders are examined, i.e. the influence of orientation and particle position on the panel is negligible. In case of the wood pellets, the calculated averaged relative standard deviation during the three test series for each of the five samples is just 0.34 %.

The error bars for both automated and manual measurements of the wood pellets (a) are larger in comparison to those of the test cylinders (b), which is mainly due to the irregular broken edges of the pellets or imprecision introduced by the manual measurements with the caliper. In case of the automated measurements, this can additionally be explained by the working principle of the digital image analysis routine based on the detected particle contours, which causes variations in the pellet lengths determined, depending on their orientations, position or on the lens' perspective onto the front surfaces. Note that

in the automatic length determination the lengths of the circular objects are taken into account with a respective average length of half of their diameter, since alternatively not considering these "circles" would lead to larger inaccuracies of the resulting lengths. The assumption taking the length of the circular objects into account with half of their diameter is justified by the manual measurements with a caliper. In the test series 1 to 5, an average length of approx. 3.23 mm was measured manually for the circular objects.
The comparison between the automatic and manual results depicts a good agreement. The averaged deviation between manual and automated measurement of pellet lengths is 0.42 %. However, it is noticeable that the manually determined average lengths are a little smaller. This is mainly the result of the accuracy of both the calibrated greyscale values as the basis of the binary images (cf. figure 5.3) and the conversion factors between pixels and millimetres. Especially through the calibration of the threshold of greyscale values, the particles' surrounding shadows can incorrectly be considered as part of the respective particles, so that the corresponding contour will be slightly too large.

To verify the accuracy of the fines determination by the suction process (including the cyclone separation) and a subsequent gravimetric measurement, a sample of test cylinders was mixed with the predefined quantity of fines (50 g, approx. 3.8 wt.-%). This mixture of particles and fines was prepared and measured five times. The initial and the resulting amount of fines c_f for every measurement is provided in figure 5.5 (a). For the five test series, the initially, manually generated fines fractions (*init*) are compared with the automatically measured amount of fines (*meas*). The averaged deviation of 2.99 % may seem comparatively high but is not surprising, as a relatively small amount of fines was added to the samples and small deviations in the gravimetrically determined mass have a significant influence.

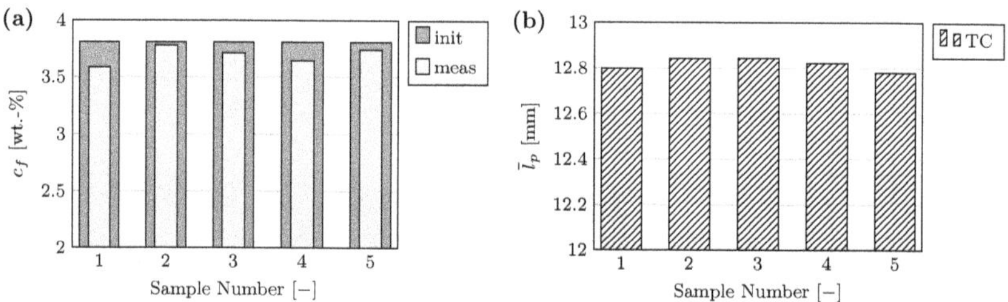

Figure 5.5: Accuracy of fines determination (a) and reproducibility of length measurements (b) [g]

Since the test cylinders had a predefined, non-changing size distribution, again, the influence of particle orientation and position on the reproducibility of the automated length measurement routine becomes obvious. The resulting average lengths are depicted in Figure 5.5 (b). In this case, the deviation only amounts 0.16 %, which is even smaller than the deviation determined before in figure 5.4 and underlines the small influence of the particle orientation and position on the LED-panel.

So far, the mechanical impact of the measuring device itself and the influence of the "circular" objects has not been determined. For determination of the degradation effect caused by the measuring system, the average pellet length of a 1 kg-sample was determined five times sequentially. In order to indicate the dependence of the comminution on the particles' durability, these test series were carried out with several pellet types. An increasing comminution with rising repetitions was found. However, since the pellet lengths are measured during typical test operation either once (simply to determine the length distribution) or at most twice (before and after, to quantify the degradation effect of a transport step), the relative deviation of 0.17 % between the first two tests with the pellets of lower durability (worst-case scenario) is acceptable. Similar to the reduced average lengths, the very small additional content of fines (in this case 0.31 wt.-% for the pellets of lower durability) has hardly any effect on the results.

Pellets that are shorter than their own diameter can randomly fall on their circular front face and remain in this position on the LED-panel. These particles would then mistakenly be detected as pellets with a length of their diameter. For this reason, these pellets (marked as "circles") are assumed to have an approximate length of half of their diameter during calculation of the average pellet length $\bar{l}_p$. Since these pellets distort the determination of the length distribution or the average length, another possibility would be to ignore those circular contoured pellets completely. For comparison, five 1 kg-samples of pellets were measured and the resulting average lengths were compared, considering and ignoring the circular objects, respectively. As expected, the average pellet lengths are generally smaller when considering the circular objects, since more comparatively short particles are taken into account. Note that the circles were already considered for the comparison between manual and automatic length measurements (cf. figure 5.4). The fact that in this case the automatically determined lengths were already larger than the manually measured values points out that neglecting the circular objects would even cause a grater gap.

5.1.2 Pneumatic conveying test facility

Design and operating principle The test rig designed allows the successive investigation into the influence of operating conditions or pipe elements on the particle flow behaviour and thus on the degradation of the products conveyed. Design and operating principle are sketched in figure 5.6. Samples of pellets (about 10 kg each) are taken from the silo and injected through a dosing system and a lock into the conveying line. The lock, consisting of four clocked and synchronised pinch valves, ensures gas tightness during continuous operation. To fill the lock with a predefined mass of pellets and simultaneously prevent particle degradation by the pinch valves, the dosing system ahead the lock contains a cone valve, which avoids particle degradation due to its operating principle (similar to the cone valve in the length measuring device, cf. figure 5.1).

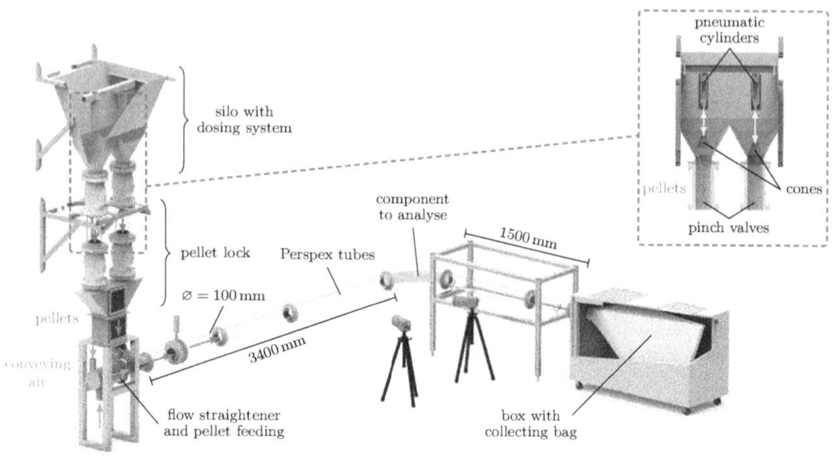

Figure 5.6: Design and working principle of the pneumatic conveying test facility [j]

The total pellet mass flow is adjusted by the timing of the pinch valves and an additional flap, which reduces the cross section downstream of the valves and ensures a controlled product flow. The conveying air, supplied by two rotating screw compressors, enters the conveying line through a flow straightener immediately upstream of the pellet feed. The air volume flow is measured by orifice metres upstream of the flow straightener. All pipes, connected by flanges, have a diameter of 100 mm, which is a typical size in practical applications, e.g. for wood pellet delivery by blowing trucks. The whole line consists of Perspex tubes, except of the specific line element under test (sketched in red). The optical access enables measuring particle movement and flow pattern at various positions.

After entering the conveying line via the feeding system, the pellets are accelerated through a 3400 mm-straight section before entering the horizontally arranged critical line component. Attached is a 1500 mm-calming or re-accelerating section. The length of the acceleration line was chosen to ensure a steady state of flow at the bend inlet. Finally, the pellets are gently separated at the end of the conveying line by a textile collecting bag, which is surrounded by a box to avoid dust release.

For quantification of the particle size reduction caused by every conveying step, the size distribution of each pellet sample is measured with the automated measuring device (section 5.1.1) before and after each conveying step.

Line components Within the scope of the current thesis, mass flows of both pellets and conveying air as well as the line component are successively varied to investigate the influence on flow and particle behaviour, but especially on pressure losses and particle degradation. Four bends of different radii as well as a 90°-Elbow as a worst-case scenario and a straight pipe section (3400 mm) for reference were applied. The bends are labelled according to their radius to diameter ratio $D_i = (r/d)$. An overview of the components varied is provided by figure 5.7. Note that depending on the component installed, the length of the conveying line ranges between 3400 mm (only acceleration line) and 5685 mm (widest bend).

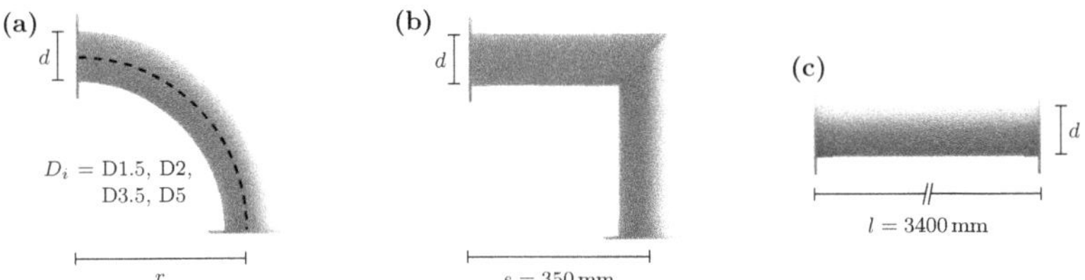

Figure 5.7: Overview and dimensioning of the pipe components with the diameter $d = 100$ mm: bends with the radius to diameter ratio $D_i = (r/d)$ (a), the 90°-Elbow (b) and the straight pipe section (c)

Particle degradation due to particle wall collisions is dependent on particle length, collision velocity and angle. For better quantitative assessment of the crushing effects caused by the bends, the maximum possible impact angles occurring within each bend section are subsequently determined using equation (5.2) [2] and compared in figure 5.8 (b).

$$\cos \varphi_i = \frac{\frac{2r}{d} - 1}{\frac{2r}{d} + 1} \qquad [-] \qquad (5.2)$$

Figure 5.8 (b) lists the resulting maximum collision angles for all components applied. It is obvious that an increase in bend radius r leads to lower collision angles. Thus, a decreasing crushing effect can be expected.

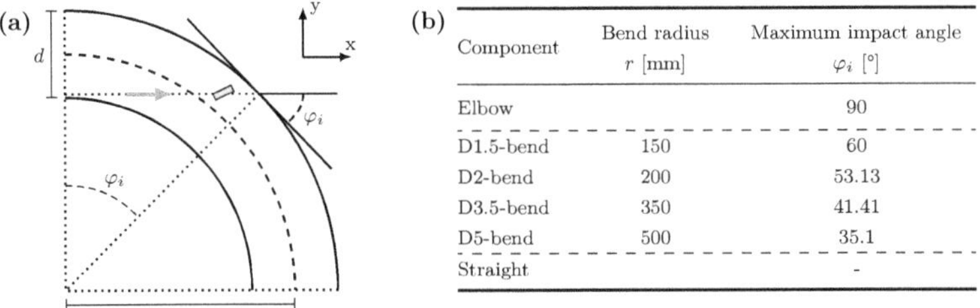

Component	Bend radius r [mm]	Maximum impact angle φ_i [°]
Elbow		90
D1.5-bend	150	60
D2-bend	200	53.13
D3.5-bend	350	41.41
D5-bend	500	35.1
Straight		-

Figure 5.8: Definition of maximum impact angle φ_i in a bend of radius r and pipe diameter d (a) [2] and calculated impact angles (b)

Pressure measurements The test rig also enables the investigation into pressure losses (always significant for dimensioning of pneumatic conveying lines). The positions of the

differential pressure transducers (halstrup-walcher P34) are schematically provided in figure 5.9. The pressure loss can be measured both over the entire conveying line Δp_0 or only over the component to be analysed Δp_1. LabVIEW®-software was used to process all data received from the pressure transducers.
Immediately ahead of the bend (at steady state particle flow), the position of the pressure transducer was varied six times circumferentially for assessing any possible influence of the particles on measurement of the pressure losses. No influences of the particles were identified. However, note that all pressure transducers were positioned at the upper pipe wall to avoid the effect of particles sliding along the pipe sheet.

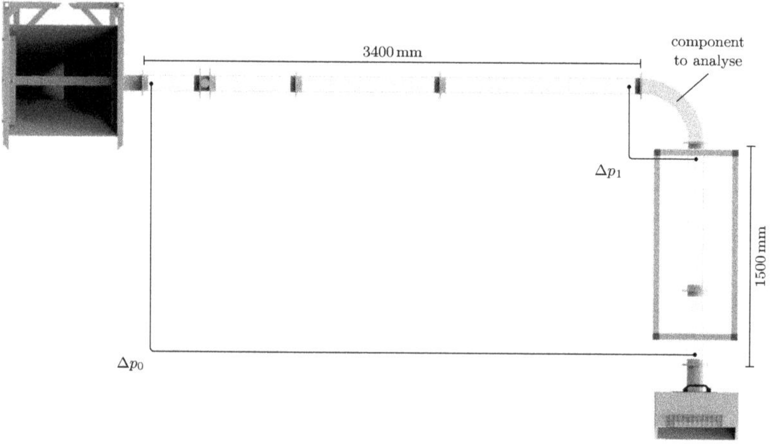

Figure 5.9: Principle of pressure measurement and positions of the pressure transducers

Conveying characteristics So-called conveying characteristics are important for better classification and comparison of the conveying line and/or the particulate bulk material conveyed. Thus, the laboratory conveying line is characterised in two steps. First, air-only pressure losses are determined for each pipe constellation investigated as a function of air mass flow (figure 5.10). Subsequently, the influence of the wood pellets on pressure loss is investigated (figure 5.11). The prevailing pressure loss is averaged over a constant period of time (20 s) and for sake of comparability, divided by the respective pipe length and denoted as $\overline{\Delta p_0}$ [$\mathrm{Pa\,m^{-1}}$] (i.e. length-specific). Note that one data point refers to a separate experiment.
The pressure drop without pellets (figure 5.10) clearly increases both with rising air mass flow and with decreasing bend radii. As expected, the straight reference pipe section and the 90°-Elbow represent excessive cases, respectively. In comparison, the prevailing pressure losses caused by the four different bends show only minor differences. Clearly visible is the quadratic relationship between air flow (i.e. air velocity) and the resulting pressure drop, in agreement with Darcy's law (equation (2.3)).

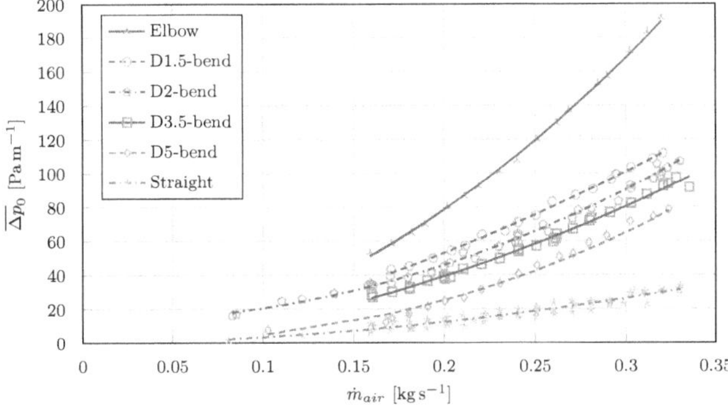

Figure 5.10: Air-only pressure drop relationship for different shape of pipe components

The pressure loss with the pellets (figure 5.11) has been measured for different air mass flows for the D2-bend. Note that for better classification, the conveying air inlet velocities are also depicted. The chart shows an increasing pressure drop for larger pellet mass flow due to larger particle quantities and thus the resulting smaller effective pipe cross sections. Figure 5.11 also shows the so-called conveying limit of the test rig. Below an air inlet velocity of 10 m s^{-1}, it is not possible to convey wood pellets continuously and reliably (e.g. without blockages or remaining particles inside the line).

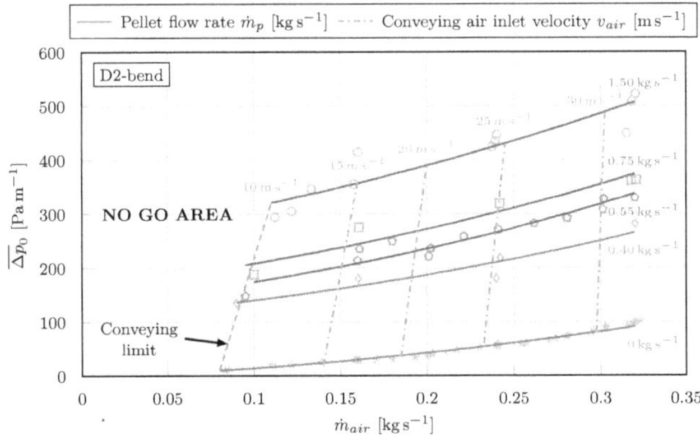

Figure 5.11: Pressure drop relationship with pellet flow rate and inlet air velocities for D2-bend pipe configuration

Reproducibility of degradation and pressure measurements For quantification and comparison of the particle degradation, a suitable indicator is required.

Following Stamboliadis [207], a crushing ratio CR is defined as

$$CR = \frac{\bar{l}_{p,0}}{\bar{l}_{p,1}}, \qquad [-] \qquad (5.3)$$

where $\bar{l}_{p,i}$ denotes the arithmetic average of the pellet length distribution of each sample before ($i = 0$) and after ($i = 1$) the conveying step. In the lower limit $CR = 1$, no degradation occurs and crushing ratios greater than one indicate progressive degradation. Contrary to simply taking the change of average pellet length as a measure for degradation, the crushing ratio considers the effect that the initial size distribution varies respectively.

To assess the repeatability, selected experiments (D2 and D5-bend) were performed five times at constant conveying conditions (see figure 5.12). The deviations of both resulting crushing ratio (a) (including the inaccuracy of the length measuring procedure) and the prevailing pressure losses (b) $\overline{\Delta p_1}$ are assessed.
Both charts indicate greater deviations for the narrower D2-bend (i.e. higher values for both crushing ratios and pressure losses) than for the D5-bend. However, the average standard deviations of the crushing ratio of ±0.36 % (D2-bend) and ±0.16 % (D5-bend) lie within reasonable limits. The average deviations of the measured specific pressure losses of ±13.7 % (D2-bend) and ±11.5 % (D5-bend) are comparatively high. From a process-technical point of view, they are within acceptable range.

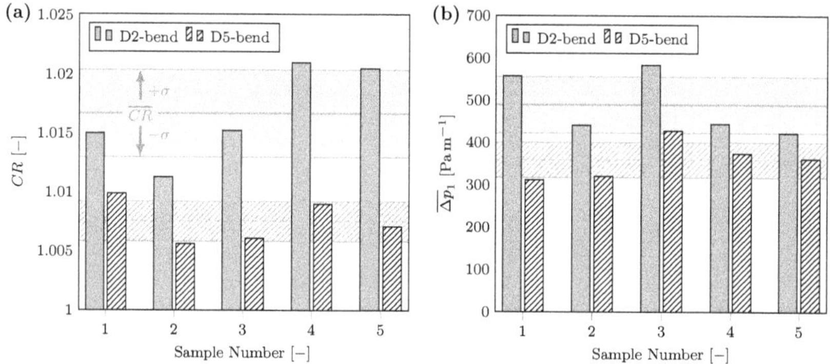

Figure 5.12: Reproducibility of the crushing ratio CR (a) and the averaged pressure loss due to the critical line component $\overline{\Delta p_1}$ (b) proved within five test series for two different pipe constellations ($\dot{m}_{air} = 0.24\,\mathrm{kg\,s^{-1}}$ and $\dot{m}_p = 0.75\,\mathrm{kg\,s^{-1}}$)

5.2 Numerical setup

5.2.1 Domain and simulation parameters

For the numerical investigation a domain consisting of DEM wall elements and the CFD mesh was set up based on the geometry of the laboratory conveying line (figure 5.13). For both the acceleration line before and the calming section after the critical line component, cross-sections are placed at different distances (l_{acc} and l_{calm}), which are applied in chapter 6 for spatial analysis, e.g. to visualise flow velocity profiles or particle distributions. Cross-sections are also provided for the bend region at different angles θ including inlet and outlet. The magnifying extract of the component to be analysed depicts the structure of the CFD mesh in the xy-plane (finer cells near the walls than in the pipe center). The pressure loss Δp is determined in terms of the difference between the area-weighted average of the local pressure at bend inlet and outlet. The inner and outer bend walls play a decisive role in the subsequent analysis of the results (chapter 6). Depending on the bend geometry applied, the number of CFD cells lies between 202 075 (straight line) and 497 117 (D5-bend).

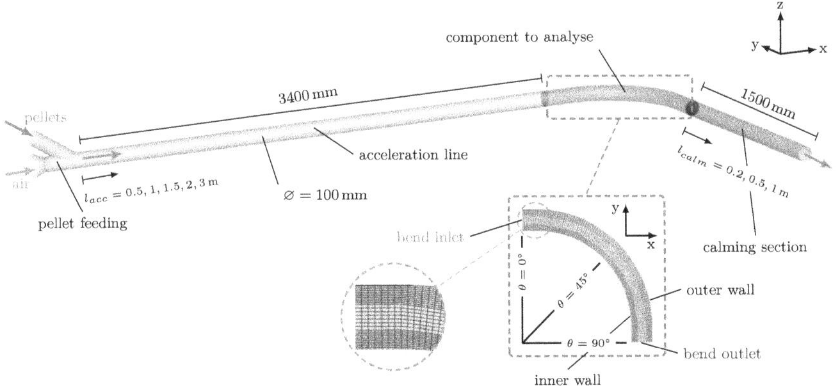

Figure 5.13: Schematic representation of the numerical domain and magnifying extract of the CFD-mesh in the bend region

In order to reduce the number of particles to a minimum and thus keeping the numerical effort within reasonable limits, individual sections were simulated separately. For this purpose, both, the particle feeding zone and the acceleration line with a length of 3400 mm were simulated in advance for each combination of air and pellet mass flows examined, under the assumption that the following bend shows no significant influence on the particle and fluid flow behaviour upstream. At the end of the acceleration line, which coincides with the inlet to the respective pipe elements of interest, all required particle properties such as velocity, orientation, angular velocity and position were stored. These serve as inlet boundary conditions for all subsequent simulation cases with varying line elements or operating conditions. Here, the critical component is simulated together with the calming section. Note that the pipe outlet is defined as pressure outlet leading into ambient pressure of 1 atm and the simulation time is 3 s (steady state flow) in all cases. For simplification, especially regarding the subsequent degradation analysis, the separation process with the textile collection bag is not considered within the numerical simulations.

The initial length distributions of the pellet samples applied are known from the experimental investigations by length measurement before each conveying step. The resulting average length distribution including the fluctuating ranges is provided in figure 5.14 (a). This distribution was applied as initial length distribution within the coupled simulations, divided in intervals of 1 mm (see figure 5.14 (b)).

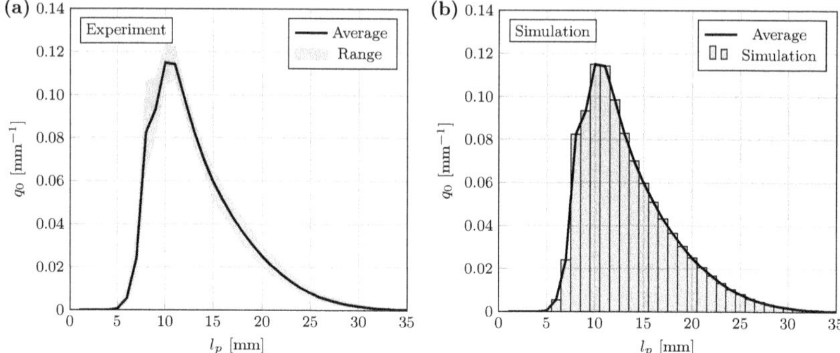

Figure 5.14: Range and initial average length distribution of the pellet bulks (a) and the numerically applied distribution (b)

The cylindrical wood pellets and the fines are approximated as spherocylinders (SCylinder) or spheres, respectively. The geometric dimensions and physical particle properties required are shown in table 5.1.

Table 5.1: Measurements, volume, mass and density of the virtual particles

	Length l_p [mm]	Diameter d_p [mm]	Sphericity ϕ [−]	Volume V_p [10^{-6} m^3]	Mass m_p [10^{-3} kg]	Density ρ_p [kg m^{-3}]
Sphere		3	1	0.01	0.02	1053
	6		0.982	0.08	0.1	
	⋮		⋮	⋮	⋮	
	10		0.943	0.2	0.21	
	⋮		⋮	⋮	⋮	
SCylinder	20	6	0.813	0.48	0.51	1053
	⋮		⋮	⋮	⋮	
	30		0.727	0.75	0.79	
	⋮		⋮	⋮	⋮	
	35		0.695	0.89	0.94	

The contact parameters applied were determined empirically or iteratively for all required material pairings between pellets and pipe materials (steel and Perspex) through drop tests, impact experiments and static or dynamic experiments of the angle of repose (see section 3.2.2). Table 5.2 gives an overview of the contact parameters applied in the current simulations.

Table 5.2: Contact parameters of wood pellets

	particle - particle	particle - steel	particle - Perspex
Coefficient of restitution e_n [−]	0.576	0.518	0.564
Collision time t_n [10^{-4} s]	5	5	5
Poisson ratio υ [−]	0.35	0.35	0.35
Static friction coefficient μ_{stat} [−]	0.657	0.564	0.518
Rolling friction coefficient μ_r [m]	0.002	0.002	0.002
Rolling friction coefficient (fines) μ_r [m]	0.9	0.9	0.9

The spring stiffnesses k^n and k^t, as well as the damping coefficient γ^n are calculated based on the time step chosen ($\Delta t_{DEM} = 5 \times 10^{-7}$ s) and the respective coefficients of restitution e_n (cf. [12, 117]). The comparatively small time step is essential for resolving the particle-particle or particle-wall contacts at high collision velocities that occur during lean phase pneumatic conveying [123, 124].
Note that in comparison to the low rolling friction coefficients for the cylindrical pellets, a very high value was chosen for the spherical fines to represent the motion-inhibiting properties.

Fluid properties The fluid properties and parameters applied for all CFD simulations are summarised in table 5.3. Constant fluid properties are assumed since an isothermal physical model is assumed (incompressible fluid) [11]. Note that both the significantly larger time step compared to DEM and the turbulence model applied are listed as well.

Table 5.3: Boundary conditions CFD

Density ρ_f [$\mathrm{kg\,m^{-3}}$]	1.255
Dynamic viscosity η_f [$\mathrm{kg\,m^{-1}\,s^{-1}}$]	1.7894×10^{-5}
Temperature T_f [K]	293.15
Timestep Δt_{CFD} [s]	5×10^{-3}
Turbulence model	k-ω-SST

5.2.2 Verification of the numerical models

First, the numerical models including corresponding model and material parameters for particle-particle and particle-wall interactions are verified. This is performed exemplary

by determination of the angle of repose for silo discharging of pellets with different fine fractions.
Subsequently, the models applied for particle-fluid interaction forces will be assessed using a single particle drop shaft scenario. The motion behaviour of a falling single particle as a function of the counteracting air flow was simulated and compared with experimental results.

5.2.2.1 Solid phase interactions

The determination of the static angle of repose α_r in dependence on the bulk's content of fines during silo discharge and the subsequent comparison between experimental and numerical data serve as verification of the numerical models implemented in DEM. Especially the modelling of the interaction between spherical (fines) and spherocylindrical particles (pellets) can be verified in this way. For this purpose, a model silo developed by Höhner [90] was applied, whose outlet can be opened abruptly.
First, the silo was filled with various layers (in total 1 kg) of pellets and fines (between 0 and 30.0 wt.-%), alternately in three layers. Figure 5.15 (a) depicts the initial experimental state of the pellet bulk for 30 wt.-% of fines. A similar procedure was applied for initialising the spherical and spherocylindrical particles in the top region of the numerical domain and then, following gravity, come to rest at the bottom of the silo. In contrast to the studies of Höhner [90], where mass flow rate and degree of discharge, flow profiles as well as particle orientation were investigated, only the resulting angles of repose of the remaining pellet bulk are applied for validation. All experiments were carried out five times in order to compensate statistical deviations by subsequent averaging of the measurement results and to ensure a better comparability of the different test series. The DEM simulations are carried out only once due to high numerical efforts.

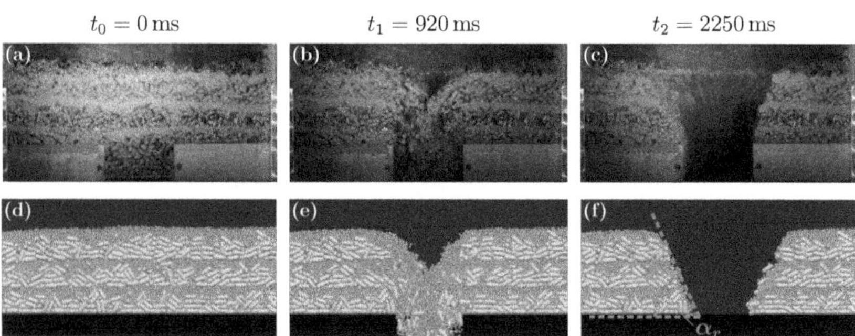

Figure 5.15: Time series of silo discharge for 30 wt.-% of fines: experiment (a)-(c), simulation (d)-(f)

The determination of the resulting angle of repose after silo discharging is carried out according to the method described by Zhou et al. [258–261]. For determination of α_r, an optical method was applied to evaluate the appropriate sequences of experimental

discharge and extracted from the simulations (cf. Höhner [90]).
Figure 5.16 compares experimental and numerical results regarding the influence of fines on the resulting angle of repose after silo discharging.

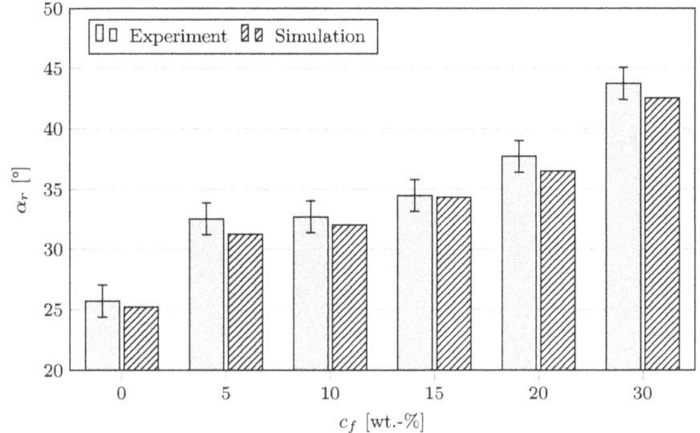

Figure 5.16: Silo repose angles dependent on the bulks' content of fines

The resulting angles of repose obviously increase with rising contents of fines - a result of the fines' motion-inhibiting properties. Basically, the experimentally determined motion behaviour is reproduced with good accuracy by the numerical models applied, including the corresponding model parameters.

5.2.2.2 Settling behaviour of single particles

To verify the DEM-CFD coupling approach, the motion of a single particle in free fall with or without counter-current fluid flow is investigated. The settling velocity and the radial displacement are numerically and experimentally determined and finally compared. In accordance to the work of Yin et al. [247] or Sørensen et al. [205], the following drop shaft test facility (figure 5.17) was built and applied (a). The corresponding numerical domain is provided in (b). The particle (or pellet) fixed in the upper part of the drop shaft by vacuum tweezers (operating on the Venturi principle) is dropped automatically. The design of the drop shaft enables air entering with a predefined mass flow rate from two sides passing through a flow straightener vertically in opposite direction to the settling particle. Two of the side walls consist of transparent Perspex, enabling the tracking of the particle's trajectory by two stereoscopically operating high-speed cameras. The tracking is performed according to the same principle mentioned in more detail in section 4.1. The cross-section depicted in (b) is coloured in z-velocity of the air flow, which clearly visualises the falling particle and thus demonstrates the particle-fluid interaction at different time steps.

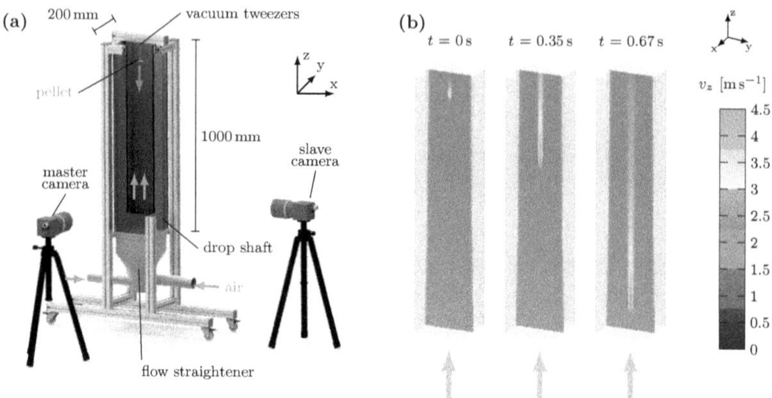

Figure 5.17: Design and working principle of the drop shaft test facility: experimental setup (a), numerical setup (b)

For better classification and to verify the experimental measurements and numerical simulations of particles in free fall, the results will first be compared with calculated values. In the following, according to [83, 212, 213], the height and velocity of a settling particle are calculated assuming that only two forces are acting on the particle:

$$\boldsymbol{F} = \boldsymbol{F}^g - \boldsymbol{F}^d \qquad [\mathrm{N}] \qquad (5.4)$$

with the gravity force $\boldsymbol{F}^g$ and the drag force $\boldsymbol{F}^d$. This balance of forces can be transformed into a time-dependent differential equation with regard to the drop height z:

$$m_p \cdot \frac{d^2\boldsymbol{z}}{dt^2} = m_p \cdot \boldsymbol{g} - \frac{1}{2} C_D \rho_f A_\perp \left(\frac{d\boldsymbol{z}}{dt}\right)^2 \qquad [\mathrm{N}] \qquad (5.5)$$

For solving this equation, the maximum settling velocity $v_{z,\infty} = v_z\,(t \to \infty)$ is calculated under the assumption that the particle does not accelerate any further ($d^2z/dt^2 = 0$). Together with the initial velocity $v_{z,0} = v_z\,(t = 0)$, the following two boundary conditions are obtained:

$$v_{z,\infty} = -\sqrt{\frac{2 m_p g_z}{C_D \rho_f A_\perp}}, \qquad v_{z,0} = 0 \qquad [\mathrm{m\,s^{-1}}] \qquad (5.6)$$

Solving the differential equation (5.5) with these boundary values, the time-related settling velocity can be calculated by

$$v_z\,(t) = -v_{z,\infty} \tanh\left(\frac{g_z}{v_{z,\infty}} t\right). \qquad [\mathrm{m\,s^{-1}}] \qquad (5.7)$$

By integrating equation (5.7), the following formula for the drop height is obtained:

$$z(t) = \frac{v_{z,\infty}(t)^2}{g_z} \ln\left(\cosh\left(\frac{g_z}{v_{z,\infty}} t\right)\right) \qquad [\mathrm{m}] \qquad (5.8)$$

These calculations are applied for comparison of the measurement and simulation results of the particles' free fall inside the drop shaft (see figure 5.19).

In addition to the settling velocity and the drop height, the radial displacement in xy-direction is applied for comparison of experiments and simulations. Figure 5.18 (a) provides the definition of the radial displacement q_r. The relevant dimensions and properties of the different particles applied for drop shaft experiments and simulations are listed in figure 5.18 (b). By selecting different particle dimensions and materials, the numerical models are validated for varying particle shapes and densities.

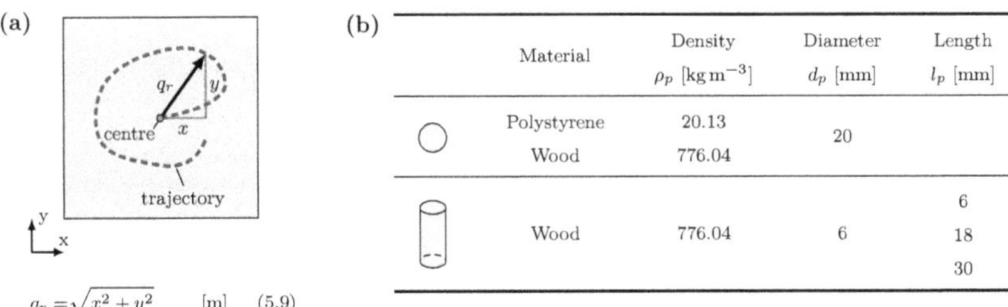

	Material	Density ρ_p [kg m^{-3}]	Diameter d_p [mm]	Length l_p [mm]
○ (sphere)	Polystyrene	20.13	20	
	Wood	776.04		
cylinder	Wood	776.04	6	6
				18
				30

$$q_r = \sqrt{x^2 + y^2} \qquad [\mathrm{m}] \qquad (5.9)$$

Figure 5.18: Topview of the drop shaft with definition of the radial displacement q_r (a) and particle's material properties of the drop shaft tests (b)

The experimentally, numerically and analytically determined settling velocities for polystyrene and wooden spherical particles of the same diameter are provided in figure 5.19.

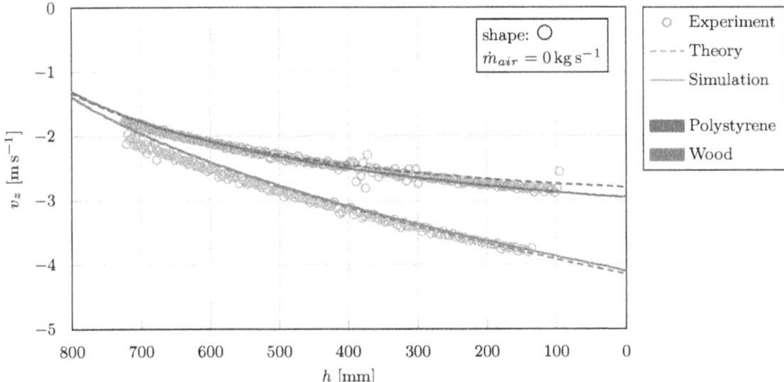

Figure 5.19: Experimentally, analytically and numerically estimated settling velocities of spherical wood and polystyrene particles

Even after a short drop height, the polystyrene particles show a considerably lower settling velocities due to their significantly smaller weight. While the experimentally determined pattern of the settling velocity as a function of drop height appears very uniform in case of the wooden particle, some deviations are visible for the polystyrene particles due to their lower density. Altogether, the experimental, numerical and theoretical data agree very well in case of a settling sphere without counter-current air flow.

The numerically determined settling velocities for cylindrical wood particles provided in figure 5.20 reflect the experimental data accurately. Simultaneously, the influences of the counter-current air flows and varying particle lengths become clearly visible. Note that the significant deviations up to half of the drop height result from the experimental tracking algorithm, which has to detect the falling particle correctly for reliable tracking. Basically, the settling velocity decreases with increasing air flow. With smaller particle length the influence of the counter-current air flow decreases. Results show two effects: On the one hand, larger particles are heavier and therefore fall faster without counter-current flow. On the other hand, they provide a larger cross-section and thus are slowed down more rapidly in case of higher counter-current air flows.

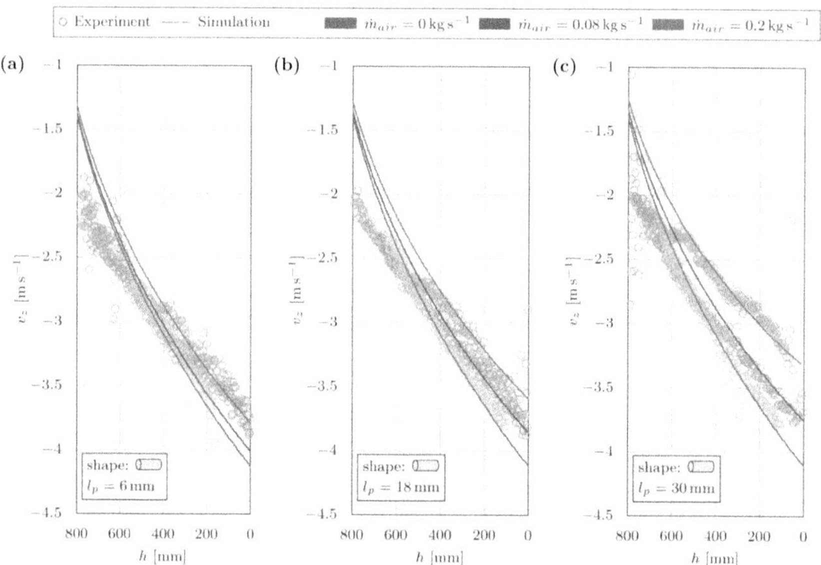

Figure 5.20: Experimentally and numerically estimated settling velocities of cylindrical wood particles of different lengths depending on different counter-current air flow rates

Figure 5.21 depicts trajectories for three cylinder lengths and counter-current air flows. Note that the coordinate systems are changed to a scale of 1:1:6 (x:y:z) for more detailed illustration of the differences between the trajectories. As expected, all three charts show that the experimentally determined trajectories show the greatest displacement from the centerline at highest air flow. In figure 5.21 (a) the displacements are smallest due to the comparatively short length (6 mm). The greatest displacements show the longest particles (figure 5.21 (c)). Although the simulated trajectories show smaller displacements than those determined experimentally, a good agreement can principally be obtained.

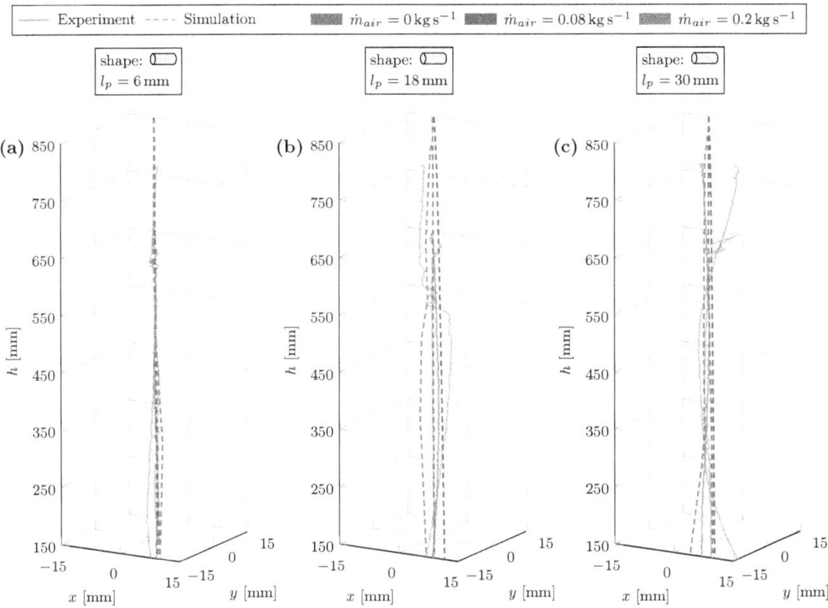

Figure 5.21: Experimentally and numerically estimated 3D-trajectories for different particle lengths and counter-current air flow rates

The differences in the particle trajectories become even clearer by the radial displacements provided in figure 5.22. Numerical results are in good agreement with the experimental data for the different particle lengths and air flows. The peaks of the experimental data (due to inaccuracies in particle tracking) are not present in the simulations.

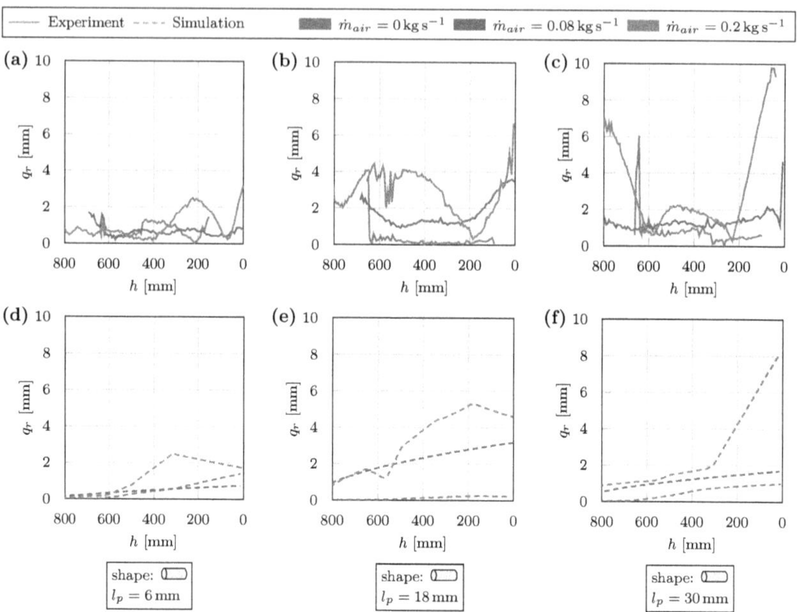

Figure 5.22: Experimentally and numerically estimated radial displacements of falling cylindrical wood particles depending on particle length and counter-current air flow rate

Generally, the comparison between numerical results and experimental data shows a good agreement for both the particle-particle and particle-wall interactions (section 5.2.2.1) as well as the particle-fluid interactions for spherical and spherocylindrical particles.

6 Conveying behaviour and particle degradation analysis

The results concentrate on particle size reduction effects of pneumatic conveying processes depending on varying operating conditions such as air and product mass flow and on bend dimensions. A second focus lies on pressure losses. Numerical results enable detailed insights into prevailing phase interactions and provide a profound explanation of the resulting particle degradation by statistical analysis of collision frequencies and velocities.
After the description in section 6.1 of the operating conditions varied, section 6.2 starts with an overview of the dependencies between particle motion, flow conditions and particle-wall interactions. Section 6.3 deals with the conveying behaviour and the resulting pressure losses in dependence on the varied operating conditions and shape of pipe components. Section 6.4 focuses on the particle size reduction with the corresponding contact frequencies and collision velocities.

6.1 Overview of investigated conditions

Table 6.1 provides an overview of the operating parameters and line components varied during experiments and numerical investigations. The operating conditions investigated numerically were chosen in accordance to the laboratory experiments. Each operating parameter, the conveying air mass flow $\dot{m}_{air}$ and the pellet mass flow $\dot{m}_p$ as well as the critical line components were varied successively. Note that for sake of simplicity, air and pellet mass flows are referred to as air and pellet flows, respectively.
A constant pellet flow of $0.75\,\mathrm{kg\,s^{-1}}$ was selected during air flow variation. The scenarios dealing with the influence of pellet mass flow are performed for an air flow of $0.24\,\mathrm{kg\,s^{-1}}$. To achieve an extended variation and thus enable a more precise analysis of the dependencies on solids loading ratio, additional investigations were carried out with an air flow of $0.32\,\mathrm{kg\,s^{-1}}$. Note that the ranges of air flow from 0.16 to $0.32\,\mathrm{kg\,s^{-1}}$ and pellet flow of 0.4 to $1.5\,\mathrm{kg\,s^{-1}}$ cover solids loading ratios between 1.25 and 6.25, mainly representing the dilute phase flow regime. The influence of the operating conditions was investigated using a D2-bend, which is frequently applied in practice for pneumatic delivery of wood pellets [49].
Experiments and simulations evaluating the influence of pipe components were performed

with several bends, where the radius-to-diameter ratio was varied in four steps (dimensioning see figure 5.7). A 90°-Elbow is applied as a worst-case scenario with respect to particle degradation. The straight pipe section provides the reference case.

Table 6.1: Selected operating conditions and pipe components

Pellet mass flow $\dot{m}_p$ [kg s^{-1}]	Air mass flow $\dot{m}_{air}$ [kg s^{-1}]								
	0.16	0.18	0.2	0.22	0.24	0.26	0.28	0.3	0.32
0.4					D2-bend				D2-bend
0.55					D2-bend				D2-bend
0.75	D2-bend	D2-bend	D2-bend	D2-bend	Elbow D1.5-bend D2-bend D3.5-bend D5-bend Straight	D2-bend	D2-bend	D2-bend	D2-bend
1.5					D2-bend				D2-bend

6.2 Dependencies among key flow features

An overview of the dependencies among the investigated key flow features shall be provided. As key flow features air flow velocity distribution, spatial particle velocity and the particle-wall interaction forces (so-called particle-wall time-averaged collision intensity, in short $WTACI$) have been selected. The $WTACI$, a parameter which characterises areas of wall erosion and particle degradation is defined as

$$WTACI = \frac{\sum_{t=0}^{T} \sum_{i=1}^{k_m} ||\boldsymbol{F}_{WC}||}{A \cdot T} \qquad [\mathrm{N\,m^{-2}\,s^{-1}}] \qquad (6.1)$$

with the respective wall element surface area A, the particle-wall collision force $\boldsymbol{F}_{WC}$ and the number of particle wall contacts k_m during the period of time considered (T).

Figure 6.1 depicts the spatial distributions of fluid velocity (a) and porosity (c) in the xy-plane of the bend region. The particle flow pattern (b) and the particle-wall interactions (d) are provided from top view in z-direction.
The influence of the particles conveyed (b) and thus the porosity distribution (c) on the fluid velocity profile (a) is clearly visible. In the region of particle rope formation near the outer bend wall, both the corresponding porosity and consequently the local fluid velocity are lower in comparison to inner bend region. Note that declined porosity gradients represent regions of frequent particle presence. The averaged particle-wall interaction forces (d) indicate the stress of individual wall regions, which is particularly relevant regarding product degradation and material wear. The example in figure 6.1 shows that the wall stresses are largest in the region of high particle occurrence (i.e. on the outer wall of the bend).

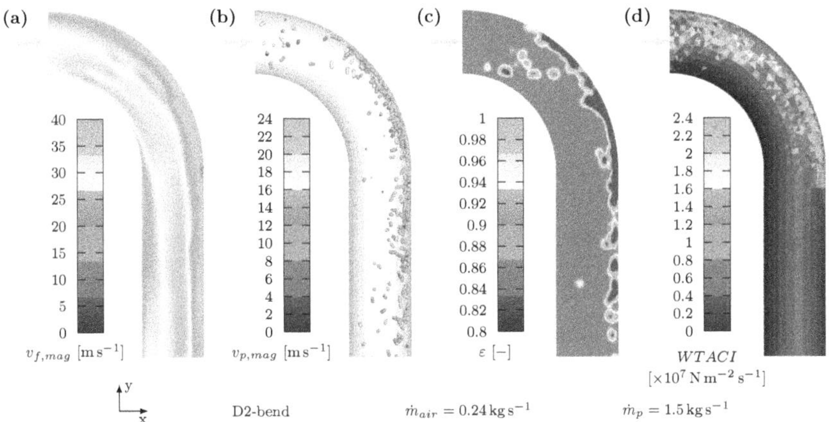

Figure 6.1: Cross-section of fluid velocity distribution (a), spatial particle velocity and flow pattern (b) and cross-section of porosity distribution (c) at random time step and time-averaged particle-wall interactions (d) for $t_{avg} = 3$ s

6.3 Conveying behaviour

6.3.1 Influence of operating conditions

Influence of air flow The spatial distributions of the time-averaged air velocities are presented in figure 6.2 for a selection of four conveying air flows in different cross-sectional views. Note that the cross-sections already defined in figure 5.13 are specified more detailed in (a). For better spatial visualisation, six cross-sections perpendicular to the pipe axis are shown in addition to the primary top view (xy-plane). The corresponding locations are indicated by dashed lines. Apart from the bend inlet and outlet, cross-sections are shown halfway along the bend (45°) and at three further locations along the calming section (l_c) reaching a distance of up to 1 m downstream of the bend outlet. This enables an additional analysis whether the fluid flow and the particle flow pattern have returned to steady state downstream the flow-disturbing pipe component. Note that in the remaining course of the current section all spatial distributions are provided similar to figure 6.2.

As expected, the average velocity level increases with rising air flows. However, the velocity gradients at the bend inlet (0°) are different. Contrary to the smallest air flow (a) where the zones of minimum velocities arise on the pipe sheet (bottom), these are located near the outer bend wall in case of the highest air flow (d). For the lowest air flow, this zone is only shifted to the outer wall at bend outlet (90°), whereas this area is already located near the outer wall halfway of the bend in both moderate cases (b) and (c). This flow behaviour already indicates prevailing rope formation, since in case of empty bends the maximum velocities are usually reached near the outer wall [151]. A typical flow profile (low velocities near walls) has already formed in all three considered cases of higher air flows at a distance of 1 m past the bend outlet, which is not the case in the scenario of the lowest flow rate. Even on the basis of the fluid velocity distributions, regions with an increased particle occurrence can be expected, which is subsequently provided in more detail in figures 6.3 and 6.4.

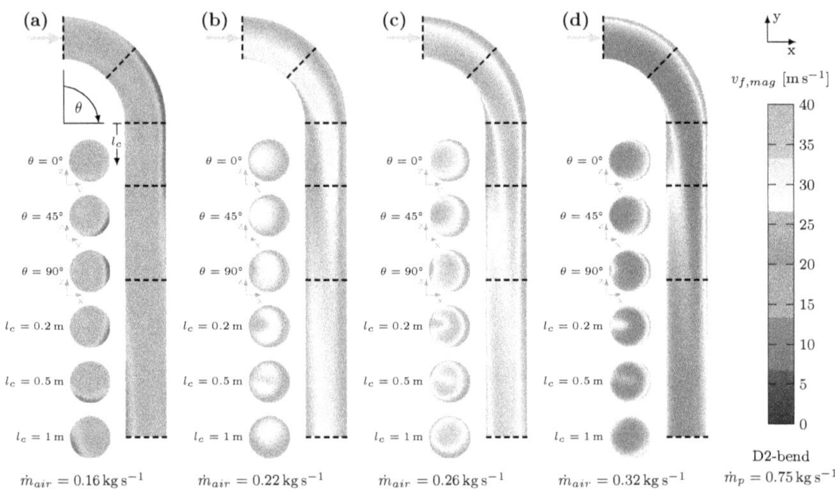

Figure 6.2: Time-averaged air velocities for varying air flows in different cross-sectional views

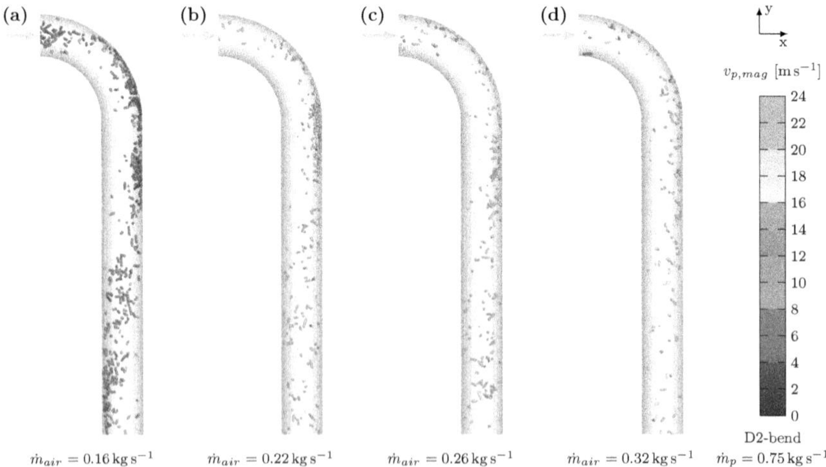

Figure 6.3: Particle flow pattern for different air mass flows at random timestep

Similar to the fluid velocities, the average particle velocity level increases with rising air flow (figure 6.3). Especially in case of the three higher air flows applied, particle velocities decrease significantly within the bend region due to direct particle-wall collisions. In accordance to the air velocity level, the particles are re-accelerated again after leaving the bend region. In all cases provided, the particles concentrate near the outer wall and form ropes when passing the bend. These streaks disperse in the straight section past the bend outlet - the sooner, the greater the re-acceleration. The scenario with the lowest air flow is an exception, since the level of particle velocity is significantly lower. Here, the re-acceleration downstream the bend outlet is comparatively low and the particles obviously still form a rope at the end of the straight pipe section.

The time-averaged porosity distributions in figure 6.4 provide a detailed spatial overview of the formed particle streaks. In the regions where ropes form and slowly disperse down stream, porosities show declined values.

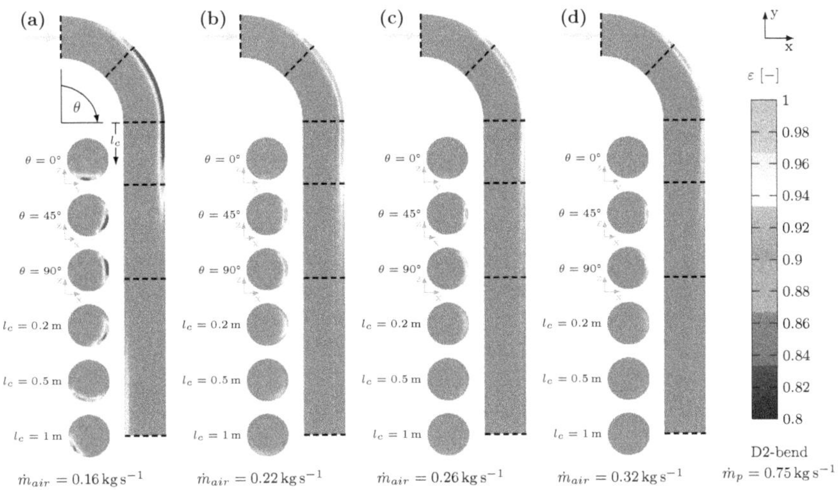

Figure 6.4: Time-averaged porosities for varying air flows in different cross-sectional views

This is clearly illustrated in case of the lowest air flow by perpendicular cross-sections along the pipe axis. The sliding of the particles along the pipe bottom is already visible at bend inlet. Here, the rope remains along the entire pipe section, but changes its spatial position. In all other scenarios, the uniform porosity level of approx. 1 indicates a homogeneous steady-state dilute phase flow regime at bend entry. In addition, the less dense ropes disperse more quickly with increasing air flows.

Since the particle velocities at bend inlet are already decisive for particle degradation, the cumulative particle inlet velocity distributions for the four varying air flows are subsequently provided in figure 6.5. Note that for all air flows applied, the averaged particle and air inlet velocities are compared subsequently in figure 6.6

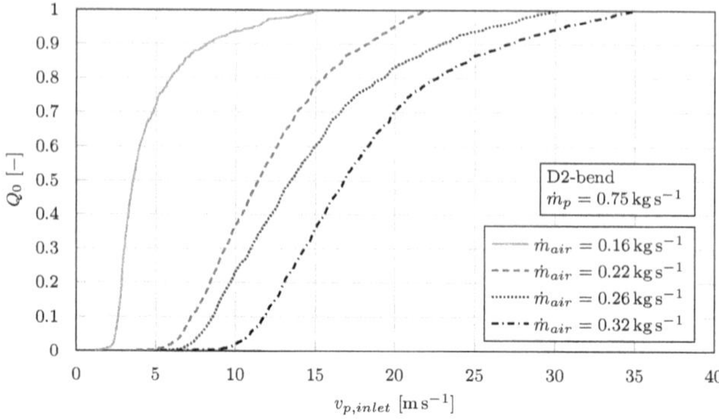

Figure 6.5: Particle velocity distributions at bend inlet for varying air flows

The shift to higher values underlines the increasing particle velocities with rising air flows. While the three scenarios of higher air flows show similar patterns, the strong increase at comparatively low inlet velocities indicates a different particle moving pattern in case of the lowest flow rate (left curve). Obviously, this is a result of the sliding motion along the pipe sheet at bend entry, already indicated by the spatial porosity distributions. These outstanding slow particle inlet velocities are also depicted in figure 6.6, where the difference between the averaged particle velocities of the two lowest air flows is large. The particle velocities are significantly lower compared to the air flow. The increase of averaged fluid velocity is of course directly proportional to the air mass flow, whereas there seems to be an increasing slip between particles and air at higher air mass flows.

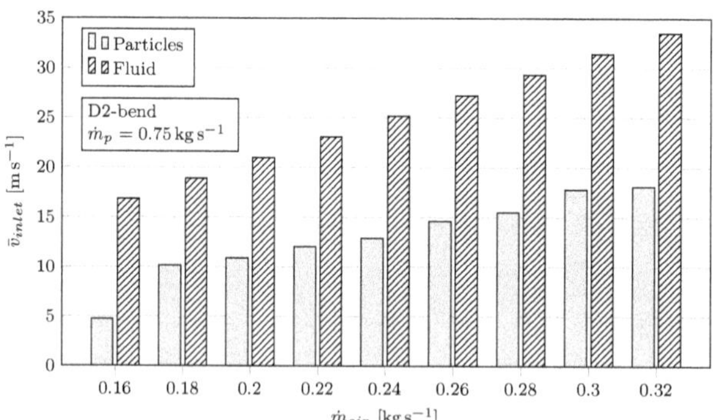

Figure 6.6: Averaged particle and fluid velocities at bend inlet for different air flows

The dependencies between air flow and resulting pressure loss are discussed next. The pressure drop was determined both experimentally and numerically, so the accuracy of the numerical models, especially concerning the inter-phase exchange of momentum, can be verified. Since the influence of different bend geometries is investigated subsequently, length-specific pressure losses are applied for uniform comparison. For sake of simplicity, time-dependent pressure drops over the bend region for 3 s of steady flow are provided in figure 6.7 exemplary for a selection of three air flow rates.

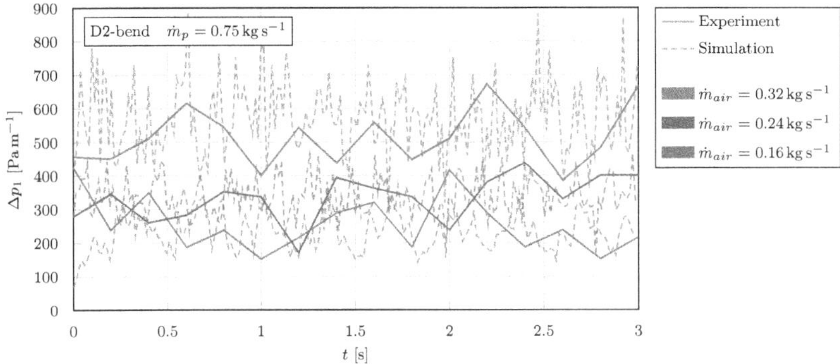

Figure 6.7: Experimentally and numerically estimated specific pressure losses over critical line component for different air flows in a period of $t = 3\,\mathrm{s}$ in steady state

Note that the intervals of numerical data storage are much shorter than the experimental measuring intervals. Consequently, the fluctuations - typical for pressure measurements in pneumatic conveying systems [214] - are different. However, the specific pressure drops determined numerically generally show acceptable agreement with the measured values. As expected, resulting pressure losses increase with rising air flows. Despite the same mass flow differences, the difference in pressure loss between the two lower air flows is smaller compared to the highest. This is caused by quadratic dependence of pressure loss on conveying velocity (cf. equations (2.3) and (2.4)).

Figure 6.8 depicts averaged specific pressure drop for all air flows.

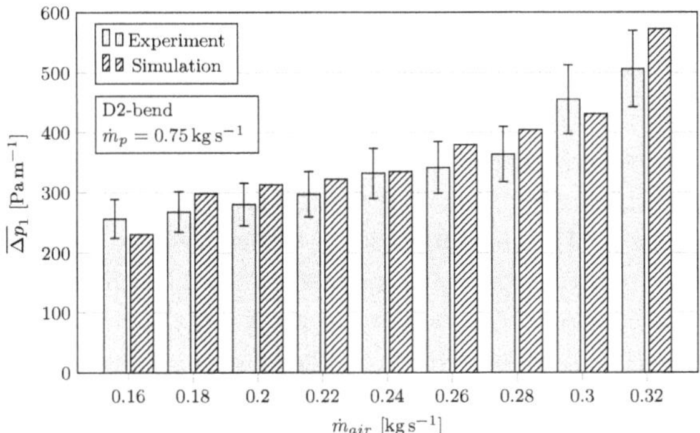

Figure 6.8: Experimentally and numerically estimated averaged specific pressure losses over critical line component for different air flows

Here, the previously mentioned non-linear increasing pressure loss with rising air flow is clearly visible. With respect to the narrow scale ($\mathrm{Pa\,m^{-1}}$), the deviations between experimental and numerical results are quite acceptable.

Impact of pellet flow Figure 6.9 depicts the effect of increasing pellet mass flow on the fluid phase based on spatial distributions of air velocity.
In comparison to the air flow, the variation of pellet flow shows only minor influence on the fluid phase for the pairs of air and pellet mass flows applied. Especially in case of the three lower pellet flows in (a) to (c), the deviations are comparatively small, partially due to low steps between the selected flow rates. Even in case of the highest pellet flow ($\dot{m}_p = 1.5\,\mathrm{kg\,s^{-1}}$ in (d)), the impact of the solid on the fluid phase becomes noticeable, especially near the outer wall in the bend region and the following straight pipe section. Nevertheless, the global velocity level remains almost the same in all scenarios presented.

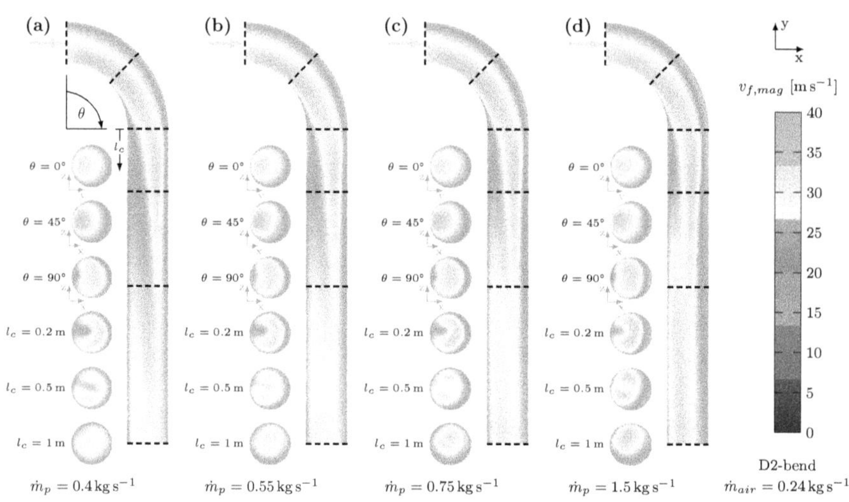

Figure 6.9: Time-averaged air velocity profiles for varying pellet flows in different cross-sectional views

Similar results provides the particle flow pattern depicted in figure 6.10. Apart from the increased particle quantity due to higher pellet mass flows, all selected scenarios show only minor differences. The absolute values of particle velocities decreases slightly with rising pellet flows. Contrary to the scenario of the lowest air flow in figure 6.3 (a), a significant rope formation outside the bend region is not visible. A detailed insight into the particle flow pattern is obtained by the spatial porosity profiles provided in figure 6.11.

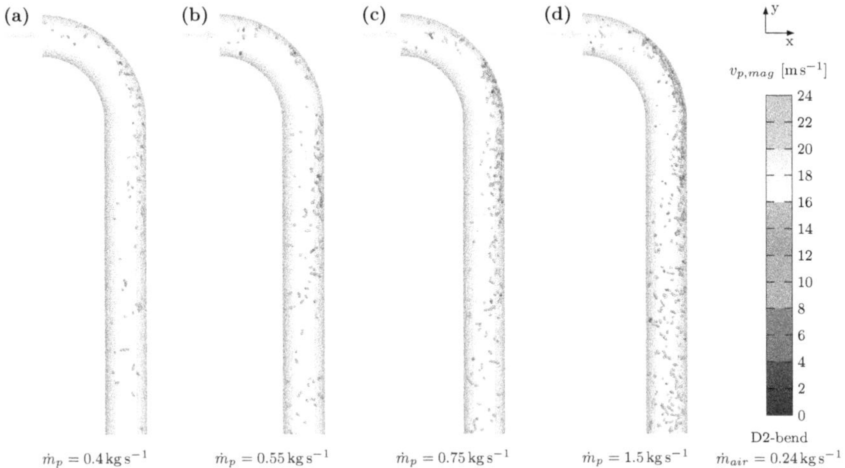

Figure 6.10: Particle flow pattern for different pellet mass flows at random timestep

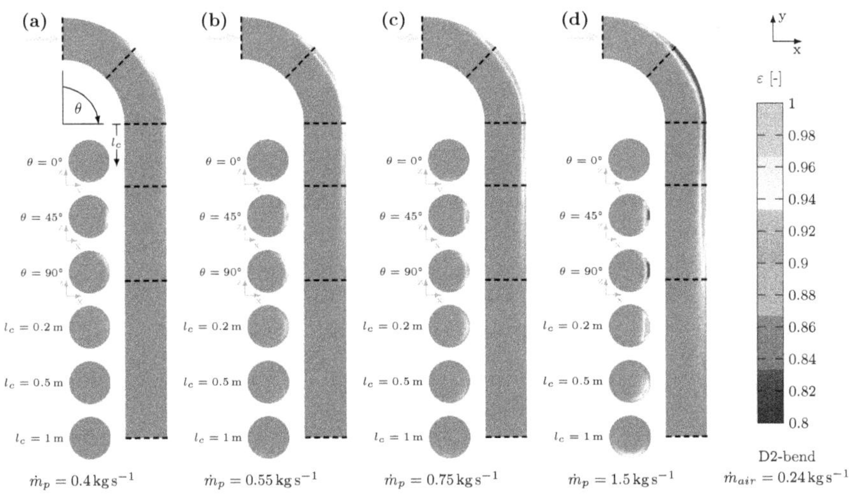

Figure 6.11: Time-averaged porosities for varying pellet flows in different cross-sectional views

Again, the uniform time-averaged porosity of approx. 1 indicates in all scenarios an initial dilute flow regime at bend entry. In every case, the particles collide with the opposite wall after entering the bend region and form a rope along the outer wall. The length of the streaks formed downstream increases with rising pellet flows, which is represented by declined porosities. After leaving the bend region, the particles are re-accelerated sufficiently and return to initial dilute flow for all three lower pellet flows. Only in case of the highest flow rate (d), an increased particle concentration at the pipe sheet at the end of the pipe section considered remains visible.

When comparing figure 6.4 (a) with the lowest air mass flow and figure 6.11 (d) with the highest pellet mass flow, although they exhibit the same solids loading ratio of approx. 4.7, a significant difference in particle motion can be detected. While figure 6.4 (a) indicated an increased particle concentration and thus rope formation (dense phase) at the pipe sheet already at bend inlet, figure 6.11 (d) shows dilute phase flow. But even within the bend and also in the straight section afterwards, there are clear differences concerning rope formation. In figure 6.4 (a), a streak is visible over the entire pipe section, whereas in case of figure 6.11 (d), the rope has almost disappeared after a distance of 1 m. This different flow behaviour underlines the classification of coarse bulk material and thus wood pellets according to Geldart into the categories B/D [189]. In contrast to fine monodisperse bulk solids such as cement, sugar or fly ash, coarse bulk materials of category B/D do not exhibit a defined and uniform fluidisation and thus conveying behaviour.

Detailed insights into the dependence of the particle velocity on product mass flow are provided in figure 6.12 by the cumulative velocity distributions at bend entry.

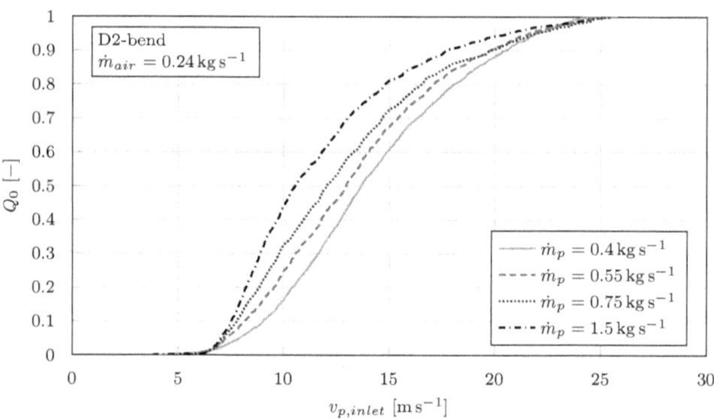

Figure 6.12: Particle velocity distributions at bend inlet for varying pellet flows

All curves show almost identical lowest and highest particle inlet velocities, approx. about 5 and 25 m s^{-1}, respectively. Here, the dependency between decreasing product mass flow and rising particle velocities, qualitatively indicated in figure 6.10, is outlined again. Whereas the plots of the lower pellet flows slightly differ from each other, the curve corresponding to highest pellet mass flow clearly indicates lower prevailing particle velocities.

This behaviour is quantitatively illustrated in figure 6.13 by averaged particle velocities at bend inlet for two different air flow levels. Therein, the particle inlet velocities are compared to the corresponding averaged air inlet velocities. Similar dependencies can be recognised for the lower (a) and the higher air velocity level (b). The averaged particle inlet velocities decrease with growing pellet flows, but are raised in magnitude for the higher air mass flow. Contrary, the averaged air inlet velocities remain almost unaffected by increasing pellet flow.

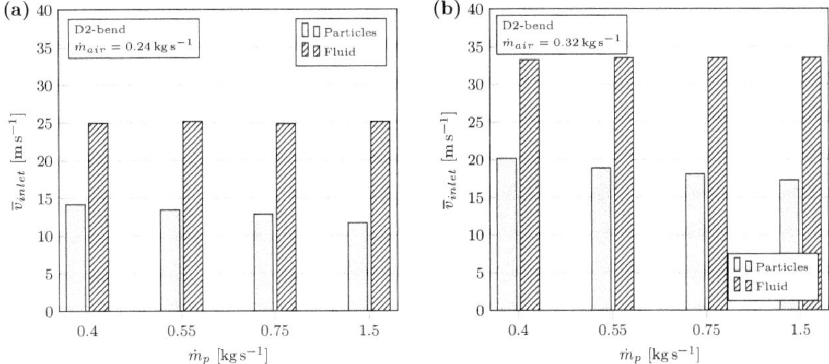

Figure 6.13: Averaged particle and fluid velocities at bend inlet for different pellet flows

The close relationship of pressure loss and particle flow is outlined in figure 6.14 by comparison of the time-dependent specific pressure drop for two different pellet flows with an empty conveying line. Apart from the general rise in pressure loss, fluctuations also increase, due to higher particle concentrations and thus more flow-disturbing irregular (non-spherical) particle motion.

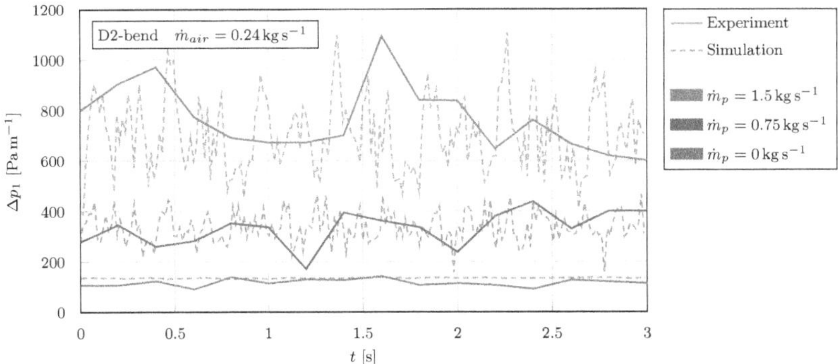

Figure 6.14: Experimentally and numerically estimated specific pressure losses over critical line component for different pellet flows in a period of $t = 3\,\mathrm{s}$ in steady state

The good agreement of experimental and numerical results is confirmed in figure 6.15 by plotting the averaged specific pressure drops in dependence of pellet flow for two different air velocity levels. Apart from the different levels, similar dependencies between rising pressure losses and pellet flows can be obtained. The most significant increase appears between the pressure drop of the empty conveying line and the lowest pellet mass flow.

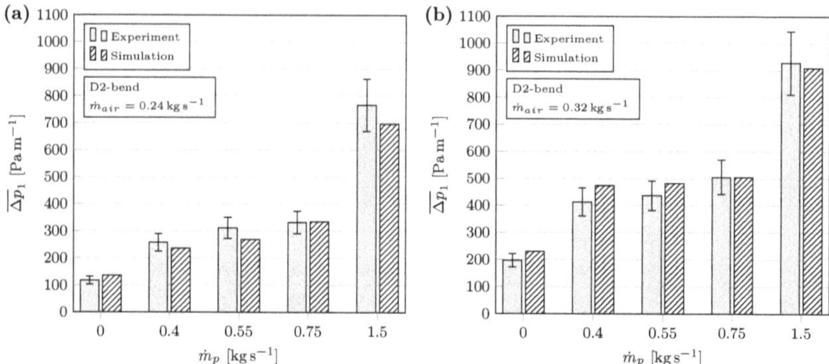

Figure 6.15: Experimentally and numerically estimated averaged specific pressure losses over critical line component for different pellet flows

Fundamentally, the increasing pressure losses with higher pellet flows result from the enhanced exchange of momentum between fluid and solid phase. Moreover, the growing particle concentration reduces the free pipe cross-section, which leads to higher local fluid velocities. This effect becomes evident in the spatial air velocity distributions in the cross-sections at the pipe outlet, e.g. in figure 6.2 (a) or figure 6.9 (d).

Effect of solids loading ratio Both air and pellet mass flows are decisive for the general particles' conveying behaviour and for pressure losses. But also the combination of air and particle flow, expressed in terms of solids loading ratio slr (cf. equation (2.2)), need to be considered. Therefore, figure 6.16 and figure 6.17 depict particle inlet velocity and pressure loss as a function of solids loading ratio for different pellet flows, respectively.

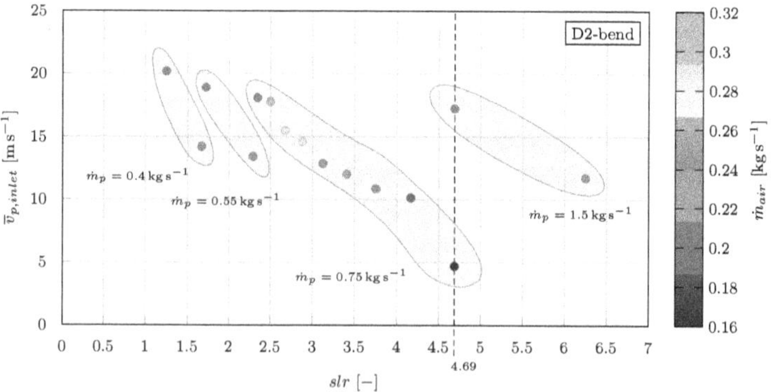

Figure 6.16: Dependence of averaged particle inlet velocities on solids loading ratio

Figure 6.16 shows declining particle velocities for increasing pellet and decreasing air flow and thus indicates the tendency of the particles' deceleration with rising solids loading ratio. Thus, it becomes obvious, that particle velocities cannot simply be deduced from *slr*. As previously mentioned, despite the same solids loading ratio of approx. 4.7, there exists a large difference between the averaged velocity for the pair of highest air and pellet flow and lowest air and second-highest pellet flow applied. This dominance of the air mass flow is underlined by the resulting almost unchanged particle velocities for different pellet flows, which is typical for dilute flow regimes [19].

The experimentally and numerically determined pressure losses, shown in figure 6.17, also do not only depend on the solids loading ratio. As before, this is indicated by the large difference between both cases of the same solids loading ratio (approx. 4.7). In addition, the quadratic relationship between pressure loss and air mass flow gets obvious, best visible for the pellet mass flow of $0.75\,\mathrm{kg\,s^{-1}}$.

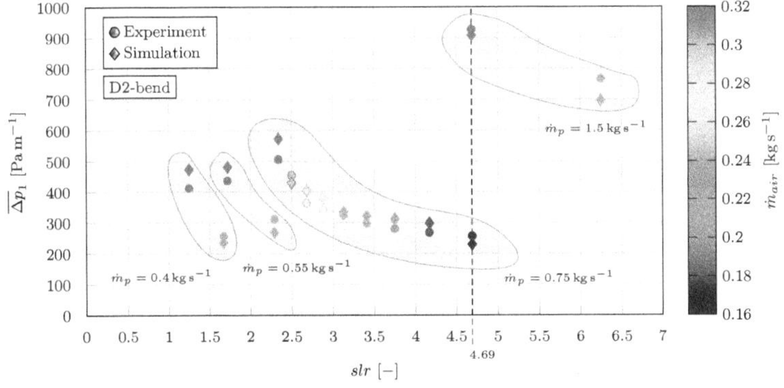

Figure 6.17: Dependence of experimentally and numerically estimated averaged specific pressure losses on solids loading ratio

6.3.2 Dependence on shape of pipe components

Figure 6.18 shows the spatial flow profile based on time-averaged velocity gradients for bends of different radii (b) to (e), together with the 90°-Elbow (a) and the straight pipe section (f). Note that the cross-sections running perpendicular to the pipe axis are arranged at different positions, depending on the appropriate bend width. For the straight section five slices are provided, each in a distance of 0.5 m. The major cross-section of the straight line does not provide the top but the side view to focus on the effect of particles possibly sliding along the pipe sheet. Each of the cross-sections at bend inlet show only minor differences in the velocity profiles, which confirms the assumption of identical inlet conditions at bend entry downstream of the former acceleration line. An exception is the inlet of the straight section, where increased velocity gradients show the influence of the nozzle and the flow straightener - positioned immediately in front of the pellet feeding (cf. figure 5.13).

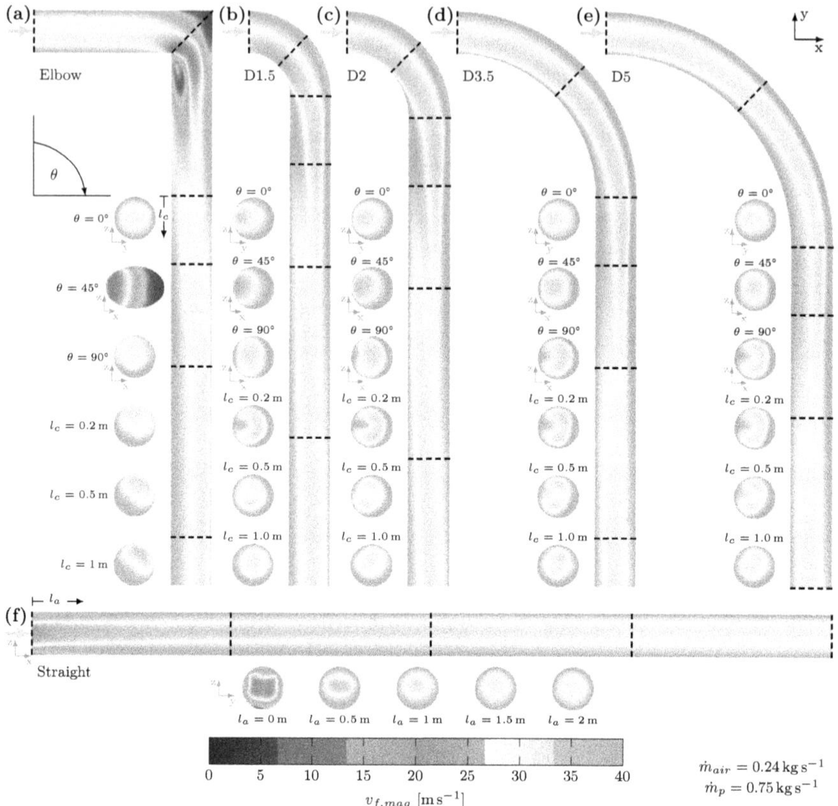

Figure 6.18: Time-averaged air velocities in dependence on shape of pipe components in different cross-sectional views

Figure 6.18 clearly indicates that the smaller the bend radius, the higher the prevailing velocity gradients. The top views illustrate a high fluid velocity level in the center of the appropriate bend and pipe section. The shape of this velocity streak is related to the extension of the bend width. The regions of lower air velocities basically result from the bend geometry (near bend outlet at the inner wall) and the presence of the particles sliding along the outer wall. The 90°-Elbow constitutes a special case, since the flow direction changes abruptly. For all bends considered (except of the 90°-Elbow), the cross-sections along the downstream calming section indicate a similar flow behaviour, reaching an almost identical pattern after a distance of 1 m.

The particle motion and velocities presented in figure 6.19 indicate that the pellets progressively slide along the outer wall with increasing bend radius. In case of the 90°-Elbow, the particles clearly hit the walls head-on instead of sliding along the outer wall, recognisable by abrupt deceleration and re-acceleration when leaving the bend region. For bends of smaller radii (e.g. D1.5 or D2), this behaviour is still visible, but declines significantly with increasing bend radius. No rope formation or sliding pattern of the particles along the pipe sheet can be detected for the straight pipe section, but further acceleration becomes evident from the increasing particle velocities.

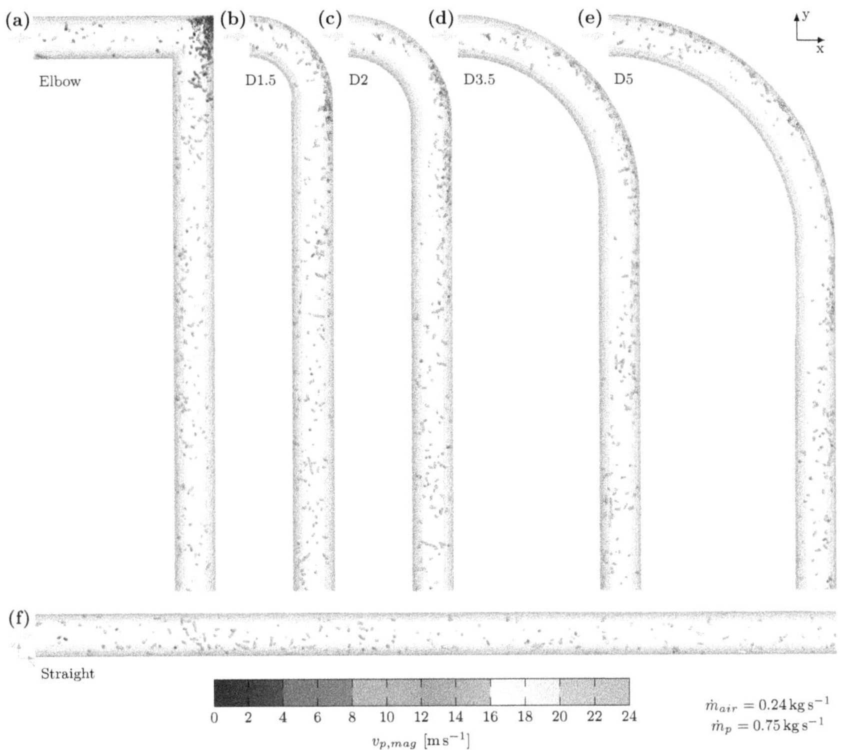

Figure 6.19: Particle flow pattern dependent on shape of pipe components at random timestep

The rope formation in individual pipe sections is clearly shown by the spatial porosity distributions in figure 6.20. In case of the 90°-Elbow, the pellets accumulate in the corner ahead of the inlet, favoured by reduced fluid velocities in this zone (cf. figure 6.18 (a)). Within the bends, the particles build ropes along the outer wall, which slightly disperses at the beginning of each calming section. The bends with larger radii (D3.5 and D5) show two zones (I and II) of reduced porosity (and hence increased particle wall interaction), which seem to be separated from each other. In contrast, the bends of smaller radii (D1.5 and D2) show just a single zone of increased particle concentration. For all components considered, the porosity profiles in the cross-sections downstream of the bend outlet indicate dispersion of the particle streaks which increases further downstream.

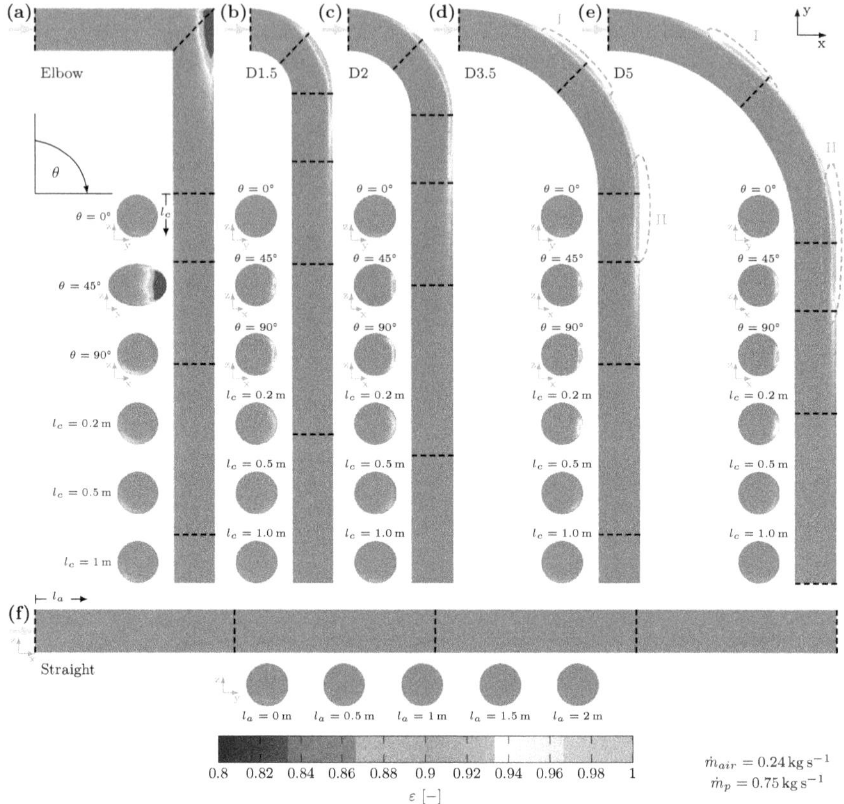

Figure 6.20: Time-averaged porosities in dependence on shape of pipe components in different cross sectional views

In figure 6.21, the effect of the component geometry on the time-dependent pressure losses determined experimentally and numerically. Note that for sake of simplicity, only four scenarios are considered, but figure 6.22 provides the averaged specific pressure losses for all components investigated. As expected, the 90°-Elbow causes much higher pressure drop than the bends and the straight pipe section. The pressure losses decline with increasing bend radius. In addition, the fluctuations decrease with larger radii, due to both the more uniform particle motion behaviour and the more homogeneous fluid flow with wider pipe bends.

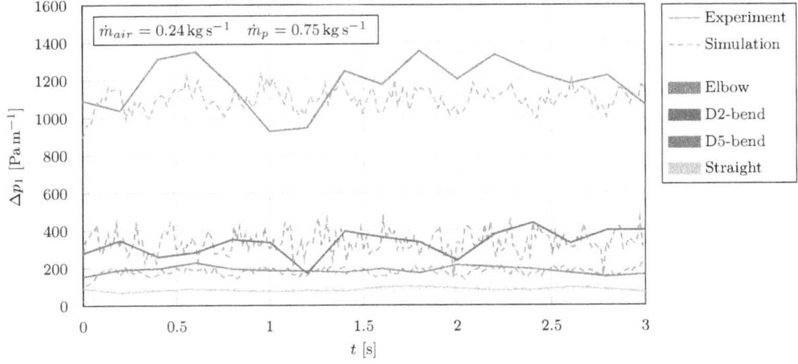

Figure 6.21: Experimentally and numerically estimated specific pressure losses over varying critical line components in a period of $t = 3\,\mathrm{s}$ in steady state

The good agreement of the numerical results with the experimental data is depicted in figure 6.22. For the bends D3.5 and D5, but especially D1.5 and D2, only slight differences in the specific pressure losses are apparent.

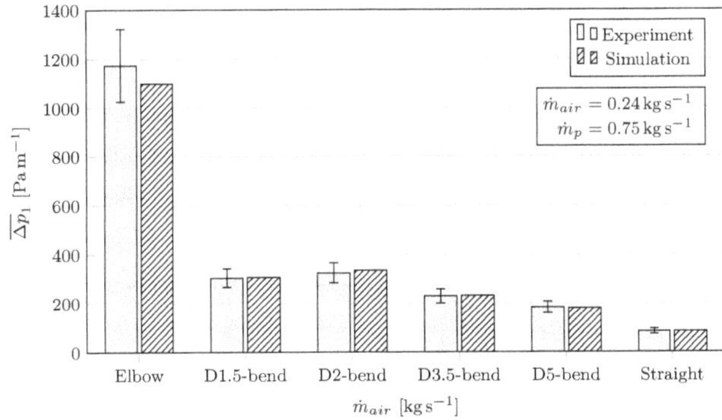

Figure 6.22: Experimentally and numerically estimated averaged specific pressure losses over varying critical line components

6.4 Particle degradation

In the following, the particle size reduction and thus fines formation during pneumatic conveying is investigated as a function of the operating conditions (air and product mass flow) and the shape of critical pipe components. Based on numerical results, the particle-wall and particle-particle collisions in terms of time-averaged particle-wall interaction forces and statistical frequency distributions of collision velocities are discussed in more detail. These provide profound insights into the resulting particle size reduction and thus consequently the formation of fines. With respect to particle degradation, numerical results are compared to experimental data to evaluate the accuracy of the degradation model developed (cf. section 4.4). Similar to section 6.3, the influences of air and pellet mass flow and consequently the solids loading ratio are discussed first. Afterwards, the dependence on the shape of pipe component is considered.

6.4.1 Assumptions and simplifications

Assumptions and simplifications were applied for numerical investigation into particle degradation during pneumatic transport, which will be described in the following. Note that the current numerical investigations explicitly focus on the mechanical influence and thus the particle breakage caused by particle-wall collisions.

- **Textile collecting bag** During experiments, the pellets conveyed were separated as carefully as possible by a textile collecting bag at the end of the laboratory conveying line (see figure 5.6). Numerically, the collecting bag's degradation effect is not considered, since the simulations only deal with the conveying line, i.e. critical line component and following straight pipe section. However, further experiments with only the straight pipe section showed minor comminution caused by the collecting bag at highest air flow rates [j].

- **Breakage by particle-wall collisions** For wood pellets, only breakage characteristics and corresponding statistical models have been investigated or developed for particle-wall collisions, respectively (see chapter 4 and [a]). It is assumed that particle-particle collision velocities are comparatively low, so they are neglected as breakage criteria.

- **Degradation by abrasion** The particle size reduction and fines formation caused by abrasion and chipping due to particle-particle or particle-wall contacts is neglected in the current work.

- **Degradation by fatigue** Material fatigue plays an important role in particle breakage during pneumatic transport, but note that it is not considered in the scope of the current work, as the focus lies on particle breakage by particle-wall collisions.

6.4.2 Influence of operating conditions

In contrast to the previous analysis of the conveying behaviour, the investigation into the effects of air and pellet mass flow is performed simultaneously, together with the directly related solids loading ratio. After the analysis of particle-wall interactions, the statistical investigation into particle-particle interactions is carried out. Based on this, the resulting degradation effect and fines formation are analysed and discussed in detail.

Particle-wall interactions The effect of varying air flow on the spatial distribution of the time-averaged particle-wall interaction force is depicted in figure 6.23, each from top and side view. Thus, both wall zones of increased particle-wall collisions and thus progressive mechanical stresses can be localised. Additionally, the plots give first indications of the associated mechanical loads on the conveyed particles.

As expected from the increasing particle velocities (cf. figure 6.6), the particle-wall interaction forces also rise with higher airflow. As a result, the area of enhanced collision forces expands, which is in any case located directly opposite to bend inlet. Noticeable is the modest wall stress in case of the smallest air flow (a) and the significant increase to the next higher one (b), which can be explained by the different particle motion behaviour (dense phase flow) and thus lower particle velocities in case of (a). With rising air flow, the sliding along the outer wall gets more intensive, clearly indicated by the side views with enlarging areas of higher interaction forces. However, the areas of intensive particle-wall contact are in all cases mainly limited to the bend region.

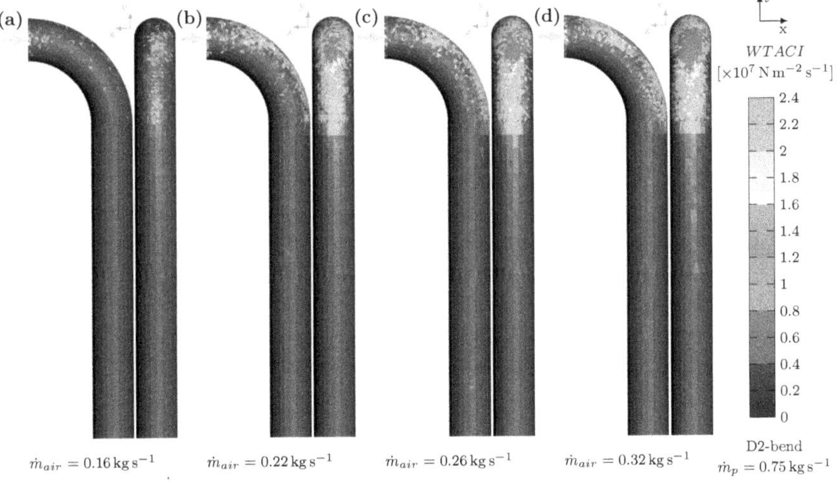

Figure 6.23: Time-averaged particle-wall interactions for different air mass flows

Figure 6.24 shows the time-averaged particle wall interactions in dependence on various pellet flows. Similar to the air flow, the strength of particle-wall interactions grows with increasing pellet flows. Since the particle velocity slightly decreases with rising pellet flow or solids loading ratio, the increasing mechanical loads of the wall elements result from rising frequency of particle-wall interactions per unit time. This reflects the reciprocal effect of increased contact frequencies (d) per unit time (at lower collision velocities) or higher interaction forces but fewer contacts (a). In case of the highest pellet flow applied, a much more extensive contact zone is observed. However, even in this case the zone of higher mechanical loads remains restricted to the bend region

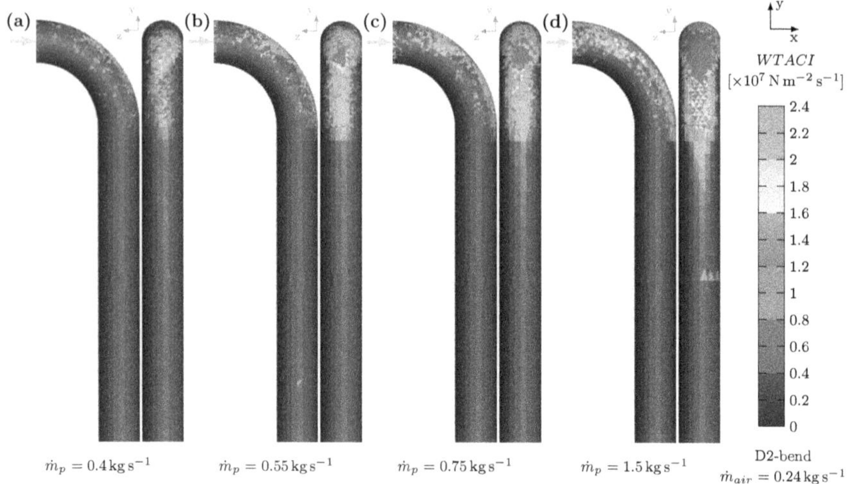

Figure 6.24: Time-averaged particle-wall interactions for different pellet mass flows

The single particle impact tests indicated that, apart from particle size, impact velocity and collision angle are decisive for the particle's degradation behaviour. The effect of impact velocity and angle can be summarised in terms of the collision velocity in normal direction to the appropriate wall element. To avoid statistical distortions due to multiple collisions at very low normal velocities (caused by particle sliding along the outer wall), figure 6.25 depicts only collision velocities in normal direction starting from a breakage-relevant absolute impact velocity v_p of $\geq 10\,\mathrm{m\,s^{-1}}$ and a collision angle φ of $\geq 20°$. The cumulative distribution in figure 6.25 illustrates the effects of varying air (a) and pellet mass flows (b) on the breakage-relevant mechanical impacts on the particles.

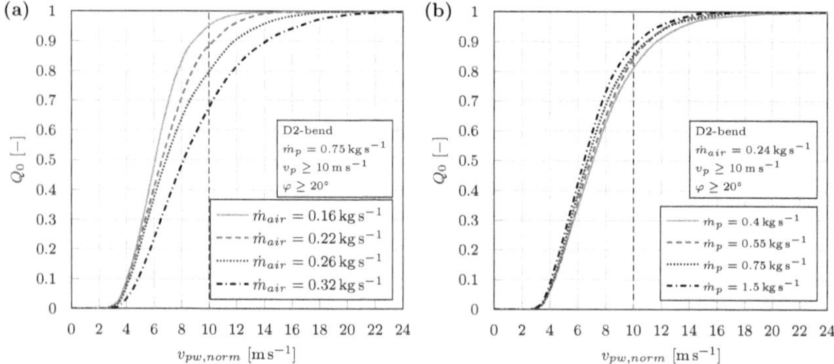

Figure 6.25: Distributions of particle-wall collision velocities in normal direction for varying air (a) and pellet flows (b)

Figure 6.25 (a) provides a clear increase of critical collision velocities with rising air flows. In the lower velocity range, the distributions show only minor differences - an effect of the restriction of the breakage-relevant collision velocities and angles. In case of the highest air flow ($0.32\,\mathrm{kg\,s^{-1}}$), about 30 % of the particle-wall contacts within the restricted range already occur at normal collision velocities of at least $10\,\mathrm{m\,s^{-1}}$. For a lower air flow of $0.16\,\mathrm{kg\,s^{-1}}$ only about 5 % of the particles exhibit a normal collision velocity about $10\,\mathrm{m\,s^{-1}}$, which indicates a lower degradation effect.
Figure 6.25 (b) shows only small differences for varying pellet flow and thus demonstrates a comparatively small influence of pellet mass flow on the breakage-relevant impact velocities. Nevertheless, it is evident that higher collision velocities occur less frequently with increasing pellet flow.

By presenting the averaged restricted "critical" impact velocities as a function of solids loading ratio, figure 6.26 confirms the previously obtained results. As higher *slr* lead to lower particle inlet velocity (compare figure 6.16), the averaged breakage-relevant normal velocities and thus the mechanical loads on the particles basically decrease with rising solids loading ratios. Nevertheless, the air flow shows stronger influence in comparison to the pellet flow for the applied range of *slr*, provided by larger gradients. However, it is not possible to automatically deduce lower impact velocities with increasing solids loading ratios (see *slr* of approx. 4.7). Obviously, the influence on the breakage-relevant collision velocities decreases with declining air flows, which can be obtained from small differences between the scenarios of lower air flow at a pellet flow of $0.75\,\mathrm{kg\,s^{-1}}$.

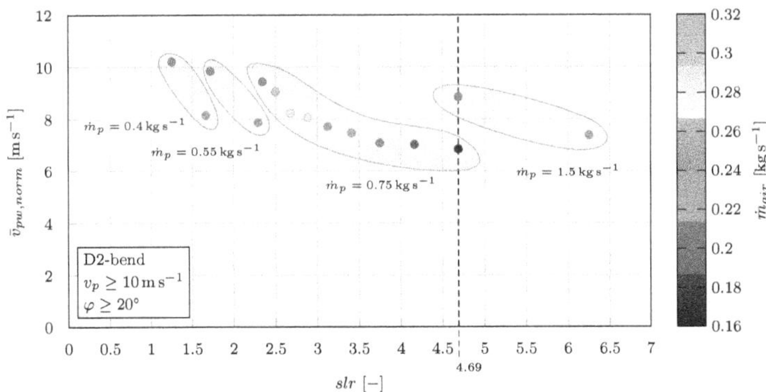

Figure 6.26: Dependence of averaged particle-wall collision velocities in normal direction on solids loading ratio

Particle-particle interactions The conveyed particles are not only exposed to mechanical loads through particle-wall but also through particle-particle interactions. Cumulative distributions of the relative particle-particle collision velocities are provided in figure 6.27 as a function of air and pellet mass flow. Note that for each particle-particle contact the corresponding relative collision velocity is considered only once. Per definition, the

mechanical loads on the contacting particles result from the change of momentum, i.e. the mechanical stress is stronger for smaller than for larger particles (see partially elastic collision between two bodies [208]). But since larger particles tend to break earlier, the relative particle-particle collision velocity is applied for statistical assessment of the mechanical loads acting on the particles through inter-particle collisions.
Similar to the particle inlet velocities, the curves in figure 6.27 (a) indicate a progressive relationship between air flow and relative collision velocities. A significant higher share of small relative collision velocities can be observed in case of the lowest air flow rate, indicated by the initially steeper gradient - a further consequence of the continuous rope formation and sliding particle motion (dense phase flow). But even in case of the highest air flow considered (i.e. highest particle velocities), nearly 80 % of the relative collision velocities are below $5\,\mathrm{m\,s^{-1}}$. Despite the duplication of the pellet mass flow and the resulting extended rope formation (cf. figure 6.11), comparatively small differences towards lower pellet flows can be observed in figure 6.27 (b). However, the inter-particle collision velocities generally decrease with increasing particle flow.

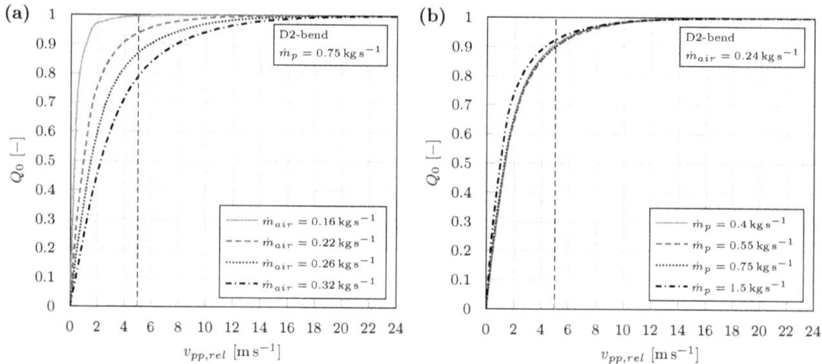

Figure 6.27: Distributions of relative particle-particle collision velocities for varying air (a) and pellet flows (b)

Figure 6.28 shows the dependence of the averaged particle-particle collision velocities on the solids loading ratio and simultaneously summarises the previous results considering all pairs of air and pellet mass flows. Additionally to the dominance of the air flow in contrast to the pellet flow, the tendency of decreasing collision velocities with rising slr becomes visible. The overall low average collision velocities imply little particle breakage due to inter-particle collisions, especially at low air velocities.

A further argument why the effect of particle-particle collision on degradation can be neglected are the elastic material properties of pellets (wood) compared to the walls (steel or Perspex).

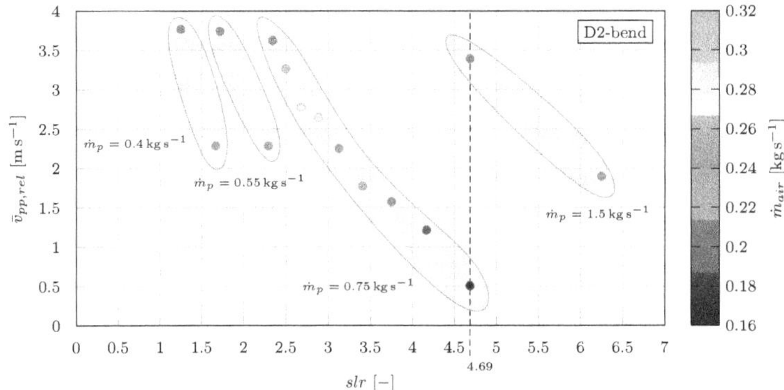

Figure 6.28: Dependence of averaged relative particle-particle collision velocities on solids loading ratio

Particle degradation and fines formation For investigating the dependence of the particles' degradation behaviour on the air and pellet mass flow, resulting cumulative size distributions, crushing ratios and contents of fines are discussed. The developed degradation model and its limits are evaluated by comparison of experimental and numerical results, especially the formation of fines, which was previously not considered during single particle impact tests. The single impact tests in chapter 4 have shown that the degradation behaviour strongly depends on the particle length. Thus, the initial length distributions must be taken into account when assessing the prevailing size reduction. The most meaningful comparison between experimentally measured and numerically determined results is therefore the crushing ratio CR (ratio between the initial and resulting average particle lengths, equation (5.3)) and the produced content of fines. Note that the length distribution of the particles are presented as well, just for completeness.

The influence of air and hence pellet mass flow on particle size reduction and fines formation is investigated in the following before analysing the effect of the solids loading ratio. Figure 6.29 shows the cumulative length distributions as a function of different air flows, determined both experimentally (a) and numerically (b). Note that the cumulative size distributions are not number- but mass-based. Otherwise comparatively high quantities of small particles caused by prevailing particle breakage would result in a misleading presentation of the respective shares. As clearly visible in each magnifying extract, the curves start from a particle length of 6 mm, which is due to the optical measuring principle, where the length of pellets detected as circles (length is less than diameter, see section 5.1.1) is averaged to half of the diameter (i.e. 3 mm). Thus, a continuous depiction of the shares of particles smaller than the actual pellet diameter is not possible. Instead, these particles are summarised by the mass fraction of particles with a length of $\leq$6 mm. Nevertheless, for determination of the average length for CR-calculation, short pellets are considered with their assumed length of 3 mm. This principle was adopted in the numerical breakage algorithm by approximating these particles by spheres of a diameter of 3 mm (see section 4.4). Note that those particles are classified as fines, which were experimentally sampled and determined gravimetrically (≤3.15 mm, see figure 5.1) or determined numerically during virtual particle fragmentation from the volume difference between resulting fragments and the corresponding mother particle (see figure 4.18). The content of fines is defined as the respective mass-related ratio between the weight of these particles and the initial sample.

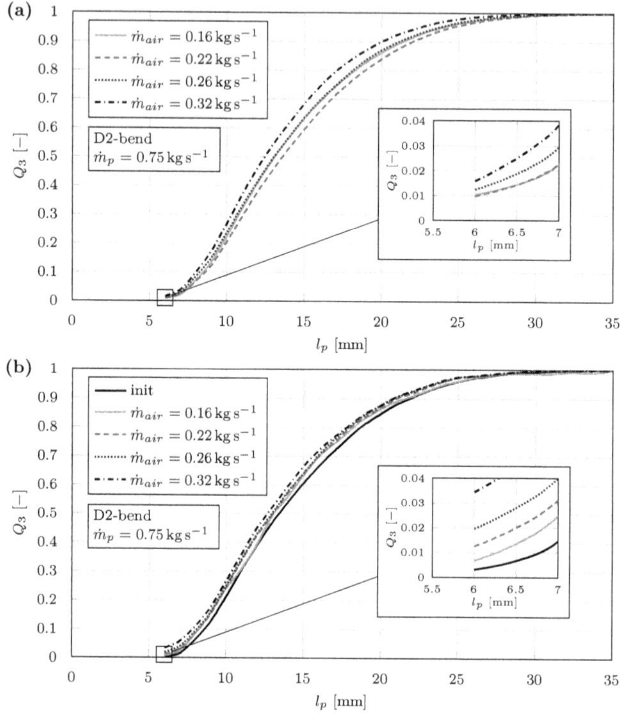

Figure 6.29: Experimentally (a) and numerically (b) determined resulting mass-related length distributions for varying air flows

As expected from the increase of particle-wall collision velocities with rising air mass flows, an increase in air flow generally results in more particle degradation. This is indicated by the shift of the distribution towards smaller particle lengths (to the left) and the higher share of particles smaller than 6 mm with increasing air flow. A tendency which is already shown by the experimental curves independent of the initial length distribution. A comparatively significant increase of small pellets with rising air flow becomes visible in the numerical distributions, especially for the highest air flow. This result is in line with figure 6.25 (a) which depicts that the particle wall collisions increase with enhanced particle-wall collision velocities.

A further comparison between experiments and simulations is provided in figure 6.30 by presenting CR (a) and the resulting fines (b), which further demonstrate the progressive relationship between particle degradation and air flow. In addition, the analogous growth of crushing ratio and the resulting contents of fines becomes obvious. Both increasing crushing ratios and fines fractions are the result of rising particle velocities and thus increasing mechanical loads acting on the particles due to particle-wall collisions. The lowest air flow considered hardly causes any comminution (CR close to 1) and comparatively little fines, whereas the highest one shows significant degradation effects ($CR > 1$). Due to measurement errors, the experimentally determined values of crushing ratio and fines do not increase uniformly or in the same ratio for each increase of air flow. However, numerical results generally agree with the experimental values, both qualitatively and quantitatively.

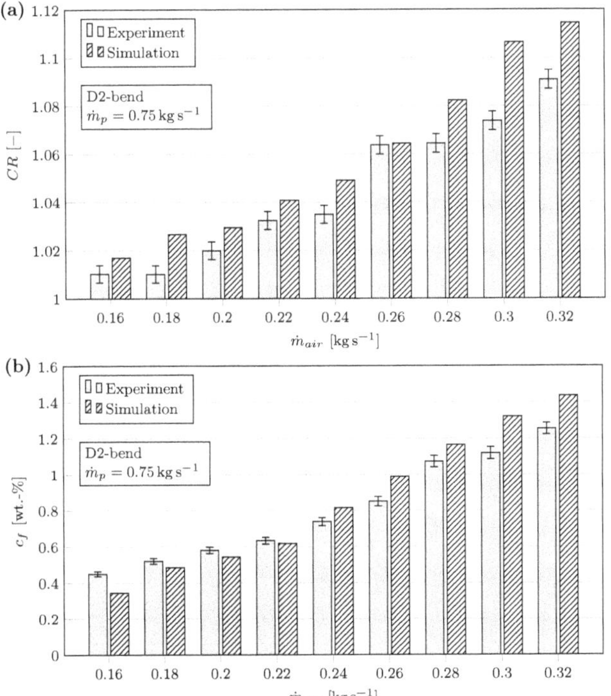

Figure 6.30: Comparison of experimentally and numerically determined crushing ratios (a) and contents of fines (b) in dependence on varying air flows

Due to the assumptions and simplifications applied, i.e. excluding particle breakage caused by fatigue, abrasion, particle-particle collisions or the textile collecting bag (see section 6.4.1), less comminution was expected within simulations. Nevertheless, numerical results show slightly increased crushing ratios and, partially, contents of fines. Indeed, the breakage probabilities are very low when the particles slide along the outer walls

(very small impact angles and normal collision velocities). But due to discretisation of the particle motion in DEM and thus the high frequency of particle-wall collisions (due to the very small time step applied), particle breakage may occur numerically, while in practice degradation is possibly caused by abrasion, not by breakage.
Furthermore, despite the very small DEM time step, high collision velocities may result in poor resolution of particle-particle or particle-wall contacts (spring-dashpot model). This possibly leads to excessive particle velocities after high-speed collisions, which result in increased breakage probabilities in case of subsequent particle-wall contact.

Figure 6.31 depicts the cumulative size distributions for different pellet mass flows at an air flow of $0.24\,\mathrm{kg\,s^{-1}}$.

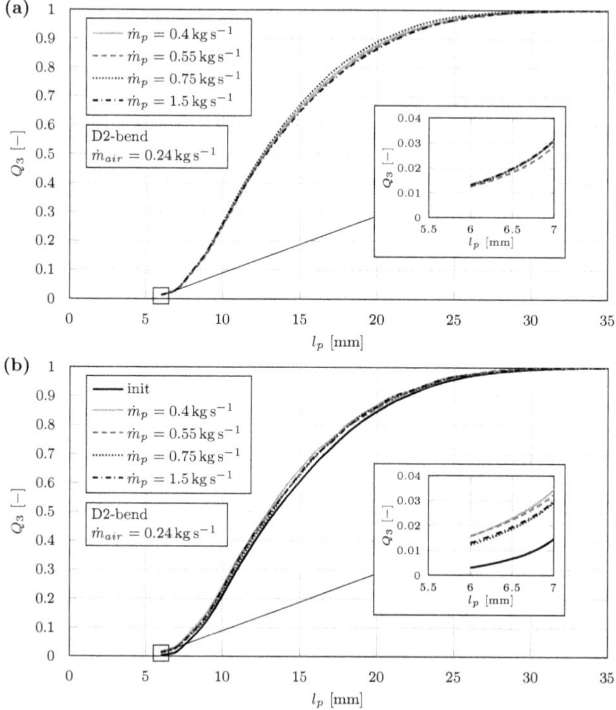

Figure 6.31: Experimentally (a) and numerically (b) determined resulting mass-related length distributions for different pellet flows

As a result of the minor influence of the pellet mass flow on the particle velocities, differences in size reduction between different pellet flows are also small. However, the trend towards higher shares of longer broken particles with increasing pellet mass flow is evident in both cases, experimentally (a) and numerically (b). Especially in the numerical case, the fractions of particles with a length $\leq 6\,\mathrm{mm}$ indicate this relationship.

Both quantitative and qualitative comparisons are provided by the crushing ratio (a) and the resulting content of fines (b) in figure 6.32. Again, the decreasing effect of increasing product mass flow on size reduction and fines formation becomes evident. Both the numerically examined crushing ratio and the fines are higher than the experimentally determined values. One reason is the enlarged number of particle-wall collisions during particle sliding along the wall due to the discretisation of particle motion and small time steps (especially in case of higher mass flows and more extensive rope formation) and thus statistically higher breakage probabilities. Even in these cases, the accuracy of the numerical results is qualitatively and quantitatively acceptable.

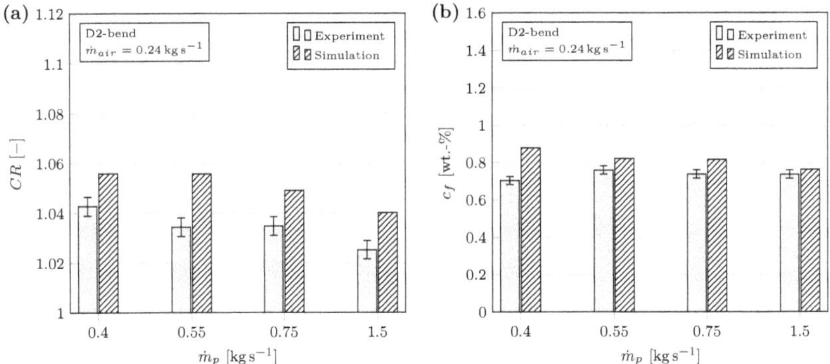

Figure 6.32: Comparison of experimentally and numerically determined crushing ratios (a) and contents of fines (b) in dependence on varying pellet flows

Figure 6.33 gives an overview of the effect of slr on the particle size reduction, taking into account all varied pairs of air and pellet mass flows, including both experimentally and numerically determined data.

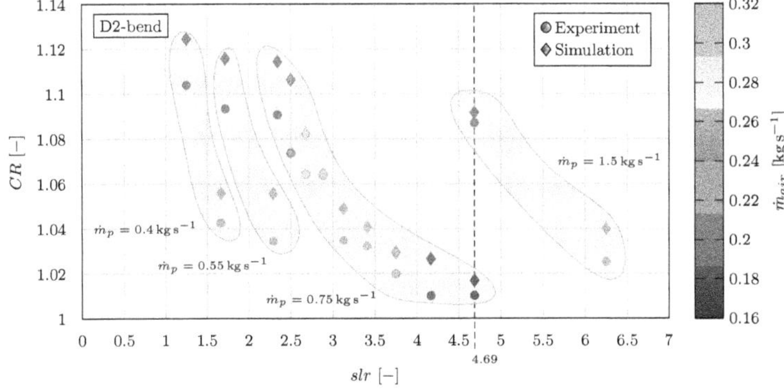

Figure 6.33: Dependence of experimentally and numerically determined crushing ratios on solids loading ratio

Note that the additional representation of the influence on the resulting fine fraction is omitted, since it correlates directly with the crushing ratio. For the considered range of solids loading ratio, a similar tendency as observed for the influence of slr on particle velocities becomes evident, i.e. less influence of pellet than of air mass flow, indicated by comparatively lower respective gradients at changing mass flows. Again and despite similar solids loading ratio (approx. 4.7), higher air flows lead to excessive size reduction. However, considering the particular air flow, a reduced degradation effect is obtained with growing solids loading ratios. This indicates that high air flows and thus velocities dominate the resulting size reduction effect in the range of slr considered.

6.4.3 Dependence on shape of pipe components

The mechanical influence on the conveyed particles and thus the degradation effect depending on the bend geometry is discussed next. Again, particle-wall interaction forces are investigated, followed by the statistical analysis of particle-wall and particle-particle collision velocities. Based on the results obtained, resulting crushing ratios and contents of fines are analysed and discussed. Numerical data are compared with experimental measurements in order to assess the accuracy of the degradation model developed.

Particle-wall interactions Figure 6.34 depicts the time-averaged particle-wall interaction forces for different bend geometries at constant solids loading ratio from top and side view. Note that only the decisively stressed wall zones are shown.
Wall regions of increased mechanical loads due to direct particle impact or enhanced sliding of the particles along the outer wall are clearly visible for all components applied, except for the straight section. The mechanical load is especially high for the region opposite the bend inlet. Whilst in case of the 90°-Elbow the region of increased mechanical stress is limited to the area opposite to the bend entry, this zone expands along the outer wall with enlarging bend radius, but with lower gradients. The sliding motion of the particles along the outer wall is particularly evident for the cases of both widest bends D3.5 and D5. Here, a second impact zone near each bend outlet becomes visible. This indicates that the particles collide with the wall in two regions instead of sliding along the outer wall. This assumption is confirmed by the declined time-averaged porosity (i.e. higher particle presence) in this area (see figure 6.20). The entire straight section (shown from side and bottom view) as well as the wall sections downstream of the 90°-Elbow remain without significant wall stresses, which again is typical for dilute phase flow.

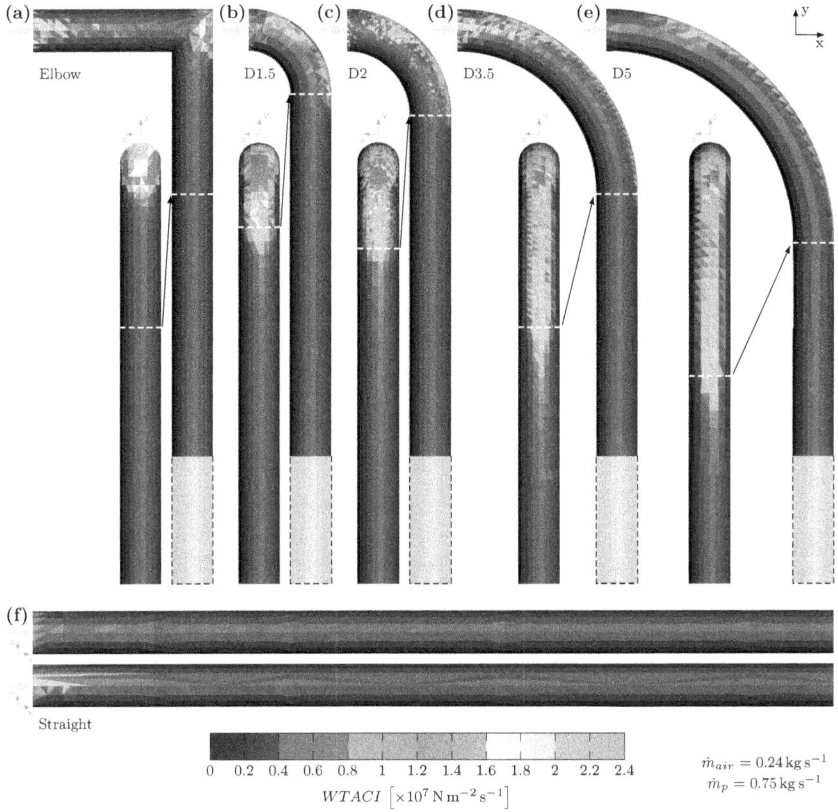

Figure 6.34: Time-averaged particle-wall interaction in dependence on the shape of pipe component

Due to its time-dependence on the collision frequencies, the time-averaged particle-wall interaction allows only limited conclusions about the resulting particle size reduction. The assessment of the collisions based on the "critical" impact velocity in normal direction provides a deeper insight into the mechanical loads acting on the particles and thus a statistical explanation for the occurring breakage effects. Both the cumulative frequency distribution (a) and the corresponding average values (b) of the occurring "critical" impact velocities in normal direction are shown in figure 6.35 for each component applied. Note that again only breakage-relevant particle-wall collisions starting at an absolute impact velocity of $\geq 10\,\mathrm{m\,s^{-1}}$ and an impact angle $\geq 20°$ were considered in order not to distort the representations by frequently occurring uncritical contacts (particle sliding along the outer wall).

The number of contacts with high and therefore breakage-relevant normal collision velocities increases with declining radius. Whilst there are only minor differences between the bends of larger radii (including the straight pipe section), the distribution of the narrowest D1.5-bend already indicates a significant increase in contacts with high degradation potential. As expected, the 90°-Elbow provides the worst-case scenario, a result of the abrupt 90°-change in flow direction. The high degradation potential of the 90°-Elbow becomes obvious when comparing the share of particles which exhibit higher values than $10\,\mathrm{m\,s^{-1}}$ with that of the D2-bend. The share is 40 % for the 90°-Elbow and only about 10 % for the D2-bend.

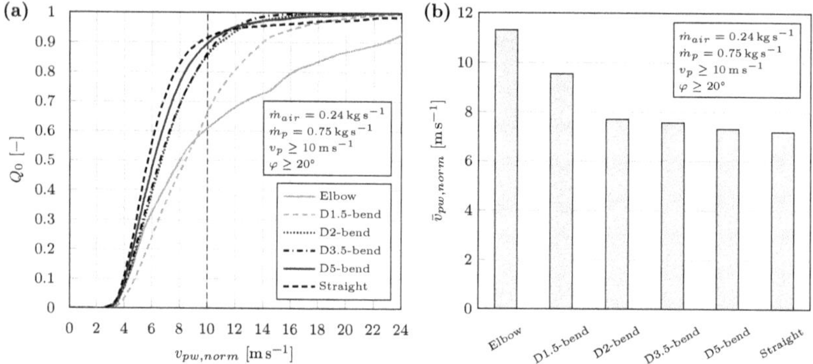

Figure 6.35: Distributions (a) and resulting averaged values (b) of particle-wall collision velocities in normal direction for varying pipe components

Similarly large or, in the case of the wider bends, smaller differences are indicated by the average collision velocities in (b). Comparatively large differences of the critical impact velocities between 90°-Elbow, D1.5-bend and the following wider bends already indicate a significant increase of particle size reduction and thus give a first outlook towards the appropriate reduction of the degradation potential with larger radius.

Particle-particle interactions The cumulative frequency distributions of the occurring inter-particle collision velocities (a) and the corresponding average values (b) provided in figure 6.36 indicate comparatively low mechanical loads acting on the particles due to inter-particle collisions, at least in case of the selected solids loading ratio.

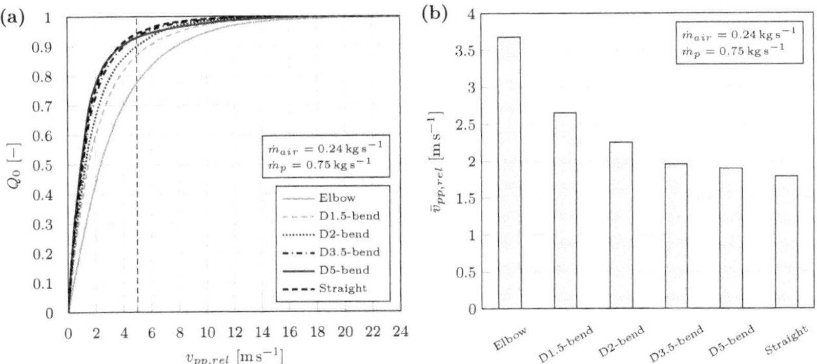

Figure 6.36: Distributions of particle-particle relative collision velocities (a) and resulting averaged values (b) for varying pipe components

The distributions show that even for the narrowest bend (D1.5) about 85 % of the particle-particle contacts occur at relative velocities of less than $5\,\mathrm{m\,s^{-1}}$, a confirmation of the simplification of neglecting particle degradation by inter-particle collisions within the implemented breakage model. An exception is the 90°-Elbow, where significantly higher collision velocities are observed, again a result of the 90°-change in flow direction, which allows direct and unimpeded inter-particle collisions (see figure 6.19). This significant difference is also shown by the average collision velocities in (b), together with the general reduction of inter-particle collision velocities with increasing bend radius up to the straight pipe section.

With increasing particle velocity levels (e.g. due to rising air flows), it is expected that the differences between the inter-particle collision velocities of the individual bend geometries will also increase. As already indicated for the D2-bend in case of the increasing air flows (figure 6.25), it can be assumed that the inter-particle collision velocities will strongly increase with rising air flows, especially in case of the 90°-Elbow. Thus, for significantly higher conveying velocities, particle degradation by particle-particle collisions should no longer be neglected for narrow bends or elbows. However, for bends whose radius enables particle sliding along the outer wall, hardly any significant particle breakage due to inter-particle collisions is to be expected.

Particle degradation and fines formation Figure 6.37 shows the experimentally (a) and numerically (b) determined cumulative length distributions. Although the experimental curves are of limited comparability due to varying initial size distributions, differences are already noticeable, especially between the straight pipe section and the bends.

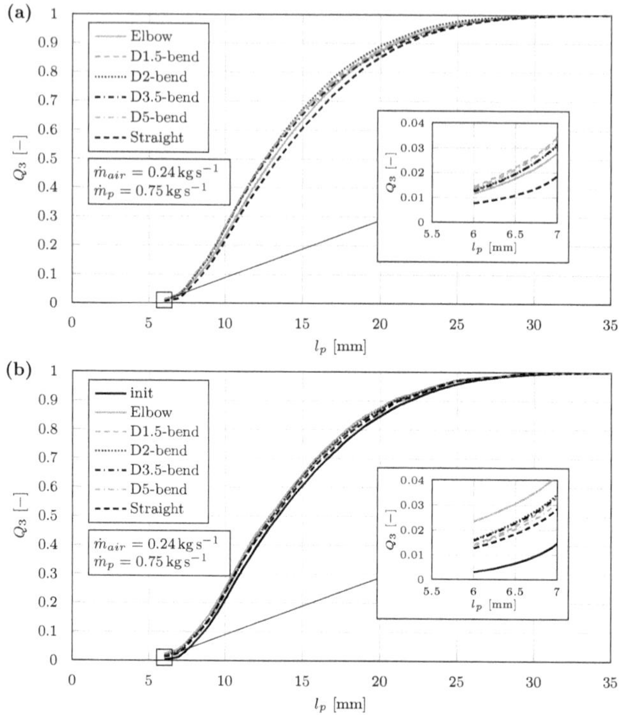

Figure 6.37: Experimentally (a) and numerically (b) determined mass-related length distributions for different pipe components

The degradation effects of the different pipe configurations become more evident in the numerical results. In addition, due to identical initial pellet lengths the data are directly comparable. In general, a significant increase of small particles ($\leq$6 mm) is obtained for all investigated pipe configurations in comparison to the initial state. The distributions and, in detail, the increasing fraction of small particles show that particle size reduction rises with smaller radius. As a worst-case scenario, the 90°-Elbow is clearly outstanding, whereas comparatively smaller differences between the individual bends and the straight pipe section are noticeable.

Figure 6.38 provides a comparison of experimentally and numerically determined size reduction effects by the crushing ratio (a) and the content of fines (b). The breakage effect of the 90°-Elbow (worst-case scenario) is significantly higher and of the straight pipe section (reference case) lower, both experimentally and numerically. This is shown by the corresponding higher or lower CR and fines, respectively. The differences between the individual bends are smaller. However, both experimental and numerical data indicate progressive degradation effect and thus more fines with decreasing radius.

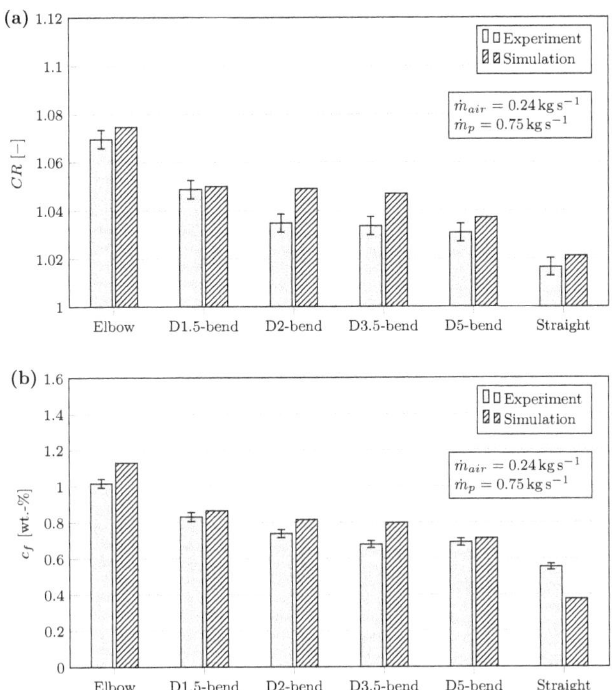

Figure 6.38: Comparison of experimentally and numerically determined crushing ratios (a) and contents of fines (b) in dependence on shape of pipe components

In summary, the numerical results show a good qualitative agreement with the experimentally determined data. As previously mentioned when investigating the influence of air and pellet mass flow (section 6.4.2), the numerically determined crushing ratios and thus the resulting fines are a little higher because of the same reasons discussed there (particle-wall and inter-particle contacts at high collision velocities may be resolved insufficiently despite the small time step; enlarged number of particle wall collisions and thus the frequent evaluation of the breakage probability due to the numerical discretisation of particle motion).

7 Summary and outlook

The consumption of wood pellets for heat and power generation in Europe increased steadily in recent years [35]. In German-speaking countries (DACH) most of the pellet production goes to domestic heating applications. Typically, pellets are delivered pneumatically into domestic storages by blowing trucks. Varying local circumstances result in a variety of conveying conditions, e.g. due to different pipe lengths or changing number of installed bends and other pipe components like reducers or couplings. Further, variations of operating conditions result from the vehicle settings set by operating staff, who, as a precaution, often select high conveying air flows to handle excessive pressure losses and avoid blockages [99].
A general problem of pneumatic conveying processes is particle size reduction, which mainly depends on the prevailing conveying conditions and line configurations. The size reduction of wood pellets needs to be avoided, since increased content of fines leads to blockages of screw conveyors or suction probes due to negative influence on the bulks motion behaviour. In addition, the storage capacity is reduced by increasing accumulation of fines, which results in higher maintenance efforts. The combustion properties of wood pellets are also impaired by smaller particles and an increased share of fines, which in turn leads to enhanced pollutant emissions.

7.1 Summary

The primary aim of the present work was to examine the influence of operating conditions (air and product mass flow) and the geometry of typical pipe components on the resulting particle size reduction and thus the formation of fines during pneumatic conveying processes. Another focus was on the resulting pressure losses. Experimental and numerical investigations were carried out using coupled DEM-CFD simulations to obtain detailed insights into flow conditions, particle motion and the mechanical loads on the conveyed pellets caused by inter-particle and particle-wall interactions. The results obtained can serve as a basis for possible measures and recommendations for material-saving pipe routing and vehicle settings.

Single particle impact tests Single particle impact tests were performed to investigate the breakage behaviour of wood pellets by particle-wall collisions. The aim was to develop empirical breakage functions for the statistical description of breakage probability

and size distribution of the resulting fragments. For this purpose, a test rig was designed, which, adapted to cylindrical particle shape, allows for a large number of repetitions under constant impact conditions. Note that only pellets of one quality were examined in the present work.
Results show that breakage characteristics and thus fines formation strongly depend on initial particle length, impact angle and collision velocity. Based on the test series, the required material-specific model parameters for the breakage functions were determined (target material: steel). Based on the developed breakage functions, a numerical degradation model (including particle breakage and fines formation) was implemented into the in-house DEM approach of LEAT for numerical determination of wood pellet degradation occurring during pneumatic conveying processes. Note that, for simplification, the developed degradation model only accounts to particle size reduction by particle-wall collisions. Other mechanical influences like abrasion, fatigue or inter-particle contacts are not considered.

Pneumatic conveying test series Experimental and numerical investigations of pneumatic conveying processes of wood pellets under variation of operating conditions and pipe components were carried out. Therefore, a pneumatic conveying line in laboratory scale was designed, which enables experimental investigation into the influence of the respective operating parameters. Apart from operating conditions like air and pellet mass flow, the bend radius was varied in four steps, supplemented by a 90°-Elbow as worst-case scenario and a straight pipe section for reference. The flow regime was chosen to cover the range between dilute phase and dense phase flow, since operating staff usually aims for high air flows and thus high conveying velocities to avoid pipe blockages. The investigations focused on the resulting particle size reduction and the occurring pressure losses. Detailed insights into resulting particle breakage was obtained from numerical results by statistical analysis of occurring mechanical loads on the conveyed particles through inter-particle and in particular particle-wall collisions.
Results indicate that particle size reduction and thus fines formation significantly depend on the conveying air flow (i.e. particle velocities) and less to pellet mass flow. With increasing air flow, collision velocities and thus particle size reduction increase, whereas for declining pellet mass flow, particle size reduction rises to a comparatively lower degree. With decreasing bend radius, the breakage effect increases due to larger collision angles. As expected, pressure losses increase with rising air or pellet mass flow and smaller bend radius.
Furthermore, it was found that due to the polydisperse size distribution and non-spherical geometry, the wood pellets cannot be clearly assigned to a flow regime on the basis of a specific solids loading ratio. Thus, the classification of wood pellets according to Geldart in class of type B/D was confirmed, since it was not possible to predict the prevailing flow regime and thus the particle velocity and pressure losses from the current solids loading ratio.
In comparison to experimental data, numerical results lead to slightly higher degradation

rates. Reasons might be particle-particle or particle-wall contacts that are not precisely resolved due to high collision velocities during pneumatic conveying and thus leading to excessive particle velocities (i.e. increase of breakage probability). Furthermore, a significant amount of particle-wall contacts arise due to the discretisation of particle motion and small time steps, especially caused by the particles' sliding along the pipe wall. Small impact angles generally lead to low breakage probabilities. But absolutely, the pellets tend to break more easily within the numerical studies due to the enlarged number of particle-wall contacts.
However, numerical results show good qualitative agreement with the experimentally determined crushing ratios and contents of fines produced. The DEM-CFD approach applied reflects the measured prevailing pressure losses with good accuracy.

The results and insights obtained show that wood pellet degradation and thus the fines formation during pneumatic transport can be reduced by suitable pipe configuration, i.e. installation of few but wide bends, avoidance of additional pipe components and choosing adequate operating conditions (reasonable ratio of air to pellet mass flow). The key findings concerning the reduction of both particle degradation as well as pressure losses are:

- Low air velocities result in low particle velocities and thus ensure less particle breakage. In addition, pressure losses decrease with decreasing air flows.
- Increased product mass flow results in lower particle velocities, which lead to less particle degradation and fines formation. However, the pressure loss increases with higher product flow.
- For cylindrical wood pellets, resulting degradation effects and pressure losses cannot be purely related to the solids loading ratio. Regarding pellet degradation and fines formation, the air flow is dominant in case of dilute phase flow regimes.
- To reduce both size reduction effects and prevailing pressure losses, bends with wide radii should be installed wherever possible.

The degradation-controlling process parameters are the particle's collision velocity and impact angle or more precisely, the particle-wall collision velocity in normal direction. The dense phase flow regime basically meets the requirement of low particle velocities and thus of less pellet breakage. However, it is hardly feasible in practice due to the large variety of line configurations and operating conditions chosen by operating staff to avoid pipe blockages during pneumatic delivery. Principally, a reasonable balance between short delivery durations (i.e. high product and air mass flows) and a gentle delivery process (lower conveying velocities) is required.
The degradation model developed allows detailed investigation into wood pellet breakage and fines formation during pneumatic conveying and simultaneously enables design and dimensioning of pipe configurations and operating conditions to prevent excessive particle size reduction and pressure losses.

7.2 Outlook

Although not presented in the current thesis, tentative analysis of bends, which are widely applied, has been initiated (figure 7.1). The detailed evaluation and optimisation of these components among low particle degradation is a further step to be taken.

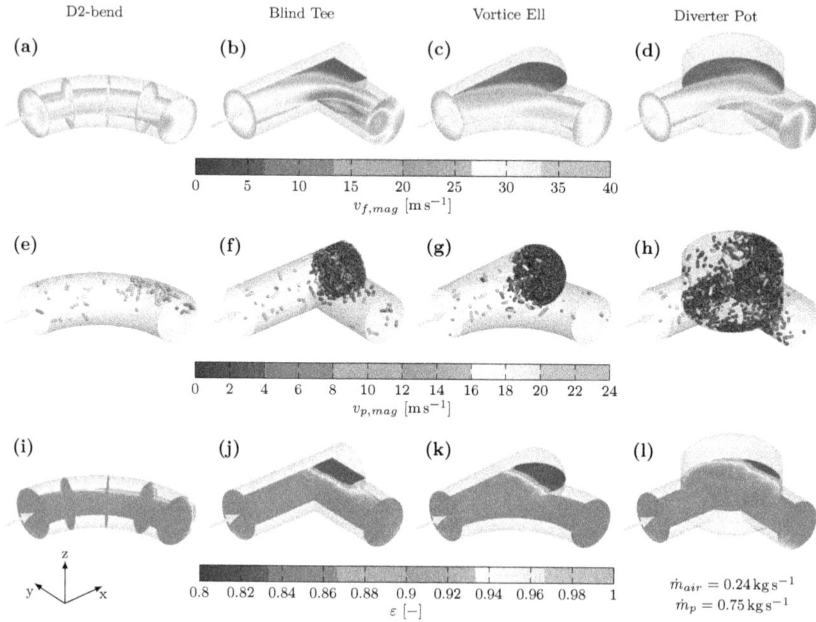

Figure 7.1: Time-averaged fluid velocities (a)-(d), particle moving pattern (e)-(h) and time-averaged porosities (i)-(l) depending on further bend geometries

Currently, the implemented numerical degradation model is limited to particle size reduction by particle-wall impacts, only for the material pairing of the present type of wood pellets and steel as target material. Nevertheless, it can easily be extended by further material-specific model parameters obtained by additional empirical impact studies, e.g. wood as target material for consideration of inter-particle collisions or HDPE, the material of impact protection mats inside domestic storages. Besides, the implemented model can be extended by additional material functions, which cover the degradation of the conveyed bulk solids by successive collisions (i.e. fatigue) or by abrasion and chipping. To avoid time-consuming DEM-CFD simulations, both the collision statistics and particle trajectories extracted in the present study can be applied as machine functions for calculating the occurring particle size reduction during pneumatic conveying in post-processing. To increase the quantitative accuracy of the numerical model, the algorithm of frequent breakage criteria examination at each particle-wall contact while particles' sliding along the wall can be optimised to avoid statistically-based excessive breakage probability. So far, the fine fraction has been approximated by spherical particles with a uniform diameter or (for small particles) recorded virtually. For more precise representation and investigation into the fines' motion behaviour, the model can be extended by an additional Euler-phase approximating the fines [197].

List of Figures

List of Tables

Bibliography

[1] **ANSYS Inc.**, *ANSYS Fluent Theory Guide*, November, Canonsburg, PA: **2013**

[2] **Aarseth K. A.**, *Attrition of feed pellets during pneumatic conveying: The influence of velocity and bend radius*, Biosystems Engineering, vol. 89, no. 2, pp. 197–213, DOI: 10.1016/j.biosystemseng.2004.06.008, **2004**

[3] **Abdulmumini M., Bradley M., Zigan S.**, *A Comparative Study of Pelletizing Process Parameters on Wood Pellet Durability*, 8th International Conference for Conveying and Handling of Particulate Solids, Tel Aviv, **2015**

[4] **Agu C. E., Tokheim L.-A., Pfeifer C., Moldestad B. M.**, *Behaviour of biomass particles in a bubbling fluidized bed: A comparison between wood pellets and wood chips*, Chemical Engineering Journal, vol. 363, no. November 2018, pp. 84–98, DOI: 10.1016/j.cej.2019.01.120, **2019**

[5] **Ai J., Chen J.-F., Rotter J. M., Ooi J. Y.**, *Assessment of rolling resistance models in discrete element simulations*, Powder Technology, vol. 206, no. 3, pp. 269–282, DOI: 10.1016/j.powtec.2010.09.030, **2011**

[6] **Alakangas E.**, *New European Pellets Standards - Introduction to standards, certification and fuel specification*, Eubionet III, vol. 4, **2011**

[7] **Antonyuk S.**, *Deformations- und Bruchverhalten von kugelförmigen Granulaten bei Druck- und Stoßbeanspruchung*, PhD Thesis, Otto-von-Guericke-University of Magdeburg, **2006**

[8] **Antonyuk S., Khanal M., Tomas J. et al.**, *Impact breakage of spherical granules: Experimental study and DEM simulation*, Chemical Engineering and Processing: Process Intensification, vol. 45, no. 10, pp. 838–856, DOI: 10.1016/j.cep.2005.12.005, **2006**

[9] **Austin L., Luckie P.**, *The estimation of non-normalized breakage distribution parameters from batch grinding tests*, Powder Technology, vol. 5, no. 5, pp. 267–271, DOI: 10.1016/0032-5910(72)80030-5, **1972**

[10] **BP**, *BP Statistical Review of World Energy 2019*, London, **2019**

[11] **Bambauer F., Wirtz S., Scherer V., Bartusch H.**, *Transient DEM-CFD simulation of solid and fluid flow in a three dimensional blast furnace model*, Powder Technology, vol. 334, pp. 53–64, DOI: 10.1016/j.powtec.2018.04.062, **2018**

[12] **Bambauer F.**, *DEM-CFD Simulation von Mehrphasensystemen mit der Anwendung auf den Hochofenprozess*, PhD Thesis, Bochum: Ruhr-University Bochum, **2019**

[13] **Beer F. P., Johnston E. R.**, *Mechanics for Engineers: Statics and Dynamics*, New York, USA: Mc-Graw-Hill, **1976**

[14] **Behera N., Agarwal V. K., Jones M. G., Williams K. C.**, *Modeling and analysis of dilute phase pneumatic conveying of fine particles*, Powder Technology, vol. 249, pp. 196–204, DOI: 10.1016/j.powtec.2013.08.014, **2013**

[15] **Biswas S., Ghoshal D.**, *Blood Cell Detection Using Thresholding Estimation Based Watershed Transformation with Sobel Filter in Frequency Domain*, Procedia Computer Science, vol. 89, pp. 651–657, DOI: 10.1016/j.procs.2016.06.029, **2016**

[16] **Boerefijn R., Ghadiri M., Salatino P.**, "Chapter 25: Attrition in Fluidised Beds", *Handbook of Powder Technology*, vol. 12, 07, pp. 1019–1053, DOI: 10.1016/S0167-3785(07)12028-5, **2007**

[17] **Bouillard J. X., Lyczkowski R. W., Gidaspow D.**, *Porosity distributions in a fluidized bed with an immersed obstacle*, AIChE Journal, vol. 35, no. 6, pp. 908–922, DOI: 10.1002/aic.690350604, **1989**

[18] **Boukherroub T., LeBel L., Lemieux S.**, *An integrated wood pellet supply chain development: Selecting among feedstock sources and a range of operating scales*, Applied Energy, vol. 198, pp. 385–400, DOI: 10.1016/j.apenergy.2016.12.013, **2017**

[19] **Bradley M. S. A.**, *Prediction of Pressure Losses in Pneumatic Conveying Pipelines*, PhD Thesis, Thames Polytechnic, **1990**

[20] **Brosch B., Scherer V., Wirtz S.**, *Simulation of municipal solid waste incineration in grate firing systems with a particle based novel Discrete Element Method*, VGB PowerTech, vol. 1, no. 2, pp. 75–83, **2014**

[21] **Brosh T., Levy A.**, *Modeling of heat transfer in pneumatic conveyer using a combined DEM-CFD numerical code*, Drying Technology, vol. 28, no. 2, pp. 155–164, DOI: 10.1080/07373930903517482, **2010**

[22] **Brosh T., Kalman H., Levy A.**, *DEM simulation of particle attrition in dilute-phase pneumatic conveying*, Granular Matter, vol. 13, no. 2, pp. 175–181, DOI: 10.1007/s10035-010-0201-z, **2011**

[23] **Brosh T., Kalman H., Levy A. et al.**, *DEM-CFD simulation of particle comminution in jet-mill*, Powder Technology, vol. 257, pp. 104–112, DOI: 10.1016/j.powtec.2014.02.043, **2014**

[24] **Buß F.**, *Geschürte Verbrennung fester Biomasse: DEM/CFD-Modellierung und experimentelle Untersuchungen*, PhD Thesis, Bochum: Ruhr-University Bochum, **2020**

[25] **CEN European Committee for Standardization**, *EN 15234-2:2012. Solid biofuels - Fuel quality assurance - Part 2: Wood pellets for non-industrial use*, **2012**

[26] **CEN European Committee for Standardization**, *ISO 17225-4:2014. Solid biofuels - Fuel specifications and classes - Part 4: Graded wood chips*, **2014**

[27] **CEN European Committee for Standardization**, *ISO 17225-5:2014. Solid biofuels - Fuel specifications and classes - Part 5: Graded firewood*, **2014**

[28] **CEN European Committee for Standardization**, *ISO 17225-6:2014. Solid biofuels - Fuel specifications and classes - Part 6: Graded non-woody pellets*, **2014**

[29] **CEN European Committee for Standardization**, *ISO 17225-7:2014. Solid biofuels - Fuel specifications and classes - Part 7: Graded non-woody briquettes*, **2014**

[30] **CEN European Committee for Standardization**, *ISO 17831-1:2015. Solid biofuels - Determination of mechanical durability of pellets and briquettes - Part 1: Pellets*, **2015**

[31] **CEN European Committee for Standardization**, *ISO 16559:2020. Solid biofuels - Terminology, definitions and descriptions*, **2020**

[32] **CEN European Committee for Standardization**, *ISO 17225-1:2020. Solid biofuels - Fuel specifications and classes - Part 1: General requirements*, **2020**

[33] **CEN European Committee for Standardization**, *ISO 17225-2:2020. Solid biofuels - Fuel specifications and classes - Part 2: Graded wood pellets*, **2020**

[34] **CEN European Committee for Standardization**, *ISO 17225-3:2020. Solid biofuels - Fuel specifications and classes - Part 3: Graded wood briquettes*, **2020**

[35] **Calderón C., Colla M., Jossart J.-M. et al.**, *Bioenergy Europe - Statistical Report 2018*, Brussels: Bioenergy Europe, **2019**

[36] **Carroll J. P., Finnan J.**, *Physical and chemical properties of pellets from energy crops and cereal straws*, Biosystems Engineering, vol. 112, no. 2, pp. 151–159, DOI: 10.1016/j.biosystemseng.2012.03.012, **2012**

[37] **Carvalho L., Wopienka E., Pointner C. et al.**, *Performance of a pellet boiler fired with agricultural fuels*, Applied Energy, vol. 104, pp. 286–296, DOI: 10.1016/j.apenergy.2012.10.058, **2013**

[38] **Chamayou A., Dodds J. A.**, "Chapter 8: Air Jet Milling", *Handbook of Powder Technology*, vol. 12, 33, pp. 421–435, DOI: 10.1016/S0167-3785(07)12011-X, **2007**

[39] **Chapelle P., Christakis N., Abou-Chakra H. et al.**, *Computational model for prediction of particle degradation during dilute-phase pneumatic conveying: modeling of dilute-phase pneumatic conveying*, Advanced Powder Technology, vol. 15, no. 1, pp. 31–49, DOI: 10.1163/15685520460740052, **2004**

[40] **Cheong Y., Reynolds G., Salman A., Hounslow M.**, *Modelling fragment size distribution using two-parameter Weibull equation*, International Journal of Mineral Processing, vol. 74, pp. 227–237, DOI: 10.1016/j.minpro.2004.07.012, **2004**

[41] **Chu K. W., Yu A. B.**, *Numerical Simulation of the Gas-Solid Flow in Three-Dimensional Pneumatic Conveying Bends*, Industrial & Engineering Chemistry Research, vol. 47, no. 18, pp. 7058–7071, DOI: 10.1021/ie800108c, **2008**

[42] **Coetzee C.**, *Review: Calibration of the discrete element method*, Powder Technology, vol. 310, pp. 104–142, DOI: 10.1016/j.powtec.2017.01.015, **2017**

[43] **Cromer A.**, *Stable solutions using the Euler approximation*, American Journal of Physics, vol. 49, no. 5, pp. 455–459, DOI: 10.1119/1.12478, **1981**

[44] **Cundall P. A., Strack O. D. L.**, *A discrete numerical model for granular assemblies*, Géotechnique, vol. 29, no. 1, pp. 47–65, DOI: 10.1680/geot.1979.29.1.47, **1979**

[45] **Dalla Valle J. M.**, *Micromeritics, the technology of fine particles*, New York, USA: Pitman Pub. Corp., **1948**

[46] **Darcy H.**, *Les fontaines publiques de la ville de Dijon*, Paris: Dalmont, **1856**

[47] **Deng T., Bradley M. S. A.**, *Determination of a particle size distribution criterion for predicting dense phase pneumatic conveying behaviour of granular and

powder materials, Powder Technology, vol. 304, pp. 32–40, DOI: 10.1016/j.powtec.2016.05.001, **2016**

[48] **Deutscher Energieholz- und Pellet-Verband e.V.**, *Lagerung von Holzpellets*, Berlin: Deutscher Energieholz- und Pellet-Verband e.V. (DEPV), **2019**

[49] **Deutsches Pelletinstitut GmbH**, *ENplus-Handbuch für Deutschland, Österreich und die Schweiz*, vol. 3, Berlin, **2015**

[50] **Dhodapkar S., Solt P., Klinzing G.**, *Understanding bends in pneumatic conveying systems*, Chemical Engineering, vol. 116, no. 4, pp. 46–52, **2009**

[51] **Di Felice R.**, *The voidage function for fluid-particle interaction systems*, International Journal of Multiphase Flow, vol. 20, no. 1, pp. 153–159, **1994**

[52] **Di Maio F. P., Di Renzo A.**, *Analytical solution for the problem of frictional-elastic collisions of spherical particles using the linear model*, Chemical Engineering Science, vol. 59, no. 16, pp. 3461–3475, DOI: 10.1016/j.ces.2004.05.014, **2004**

[53] **Di Renzo A., Di Maio F. P.**, *Comparison of contact-force models for the simulation of collisions in DEM-based granular flow codes*, Chemical Engineering Science, vol. 59, no. 3, pp. 525–541, DOI: 10.1016/j.ces.2003.09.037, **2004**

[54] **Döring S.**, *Pellets als Energieträger*, Springer Verlag Berlin Heidelberg, DOI: 10.1007/978-3-642-01624-0, **2011**

[55] **Dosta M., Antonyuk S., Heinrich S.**, *Multiscale Simulation of Agglomerate Breakage in Fluidized Beds*, Industrial & Engineering Chemistry Research, vol. 52, no. 33, pp. 11275–11281, DOI: 10.1021/ie400244x, **2013**

[56] **Du J., Hu G., Fang Z., Gui W.**, *Accelerating CFD-DEM simulation of dilute pneumatic conveying with bends*, International Journal of Fluid Machinery and Systems, vol. 8, no. 2, pp. 84–93, DOI: 10.5293/IJFMS.2015.8.2.084, **2015**

[57] **Ebrahimi M., Crapper M., Ooi J. Y.**, *Numerical and experimental study of horizontal pneumatic transportation of spherical and low-aspect-ratio cylindrical particles*, Powder Technology, vol. 293, pp. 48–59, DOI: 10.1016/j.powtec.2015.12.019, **2016**

[58] **Elskamp F., Kruggel-Emden H., Hennig M., Teipel U.**, *A strategy to determine DEM parameters for spherical and non-spherical particles*, Granular Matter, vol. 19, no. 3, DOI: 10.1007/s10035-017-0710-0, **2017**

[59] **Eltrop L., Hartmann H., Härdtlein M. et al.**, *Leitfaden Feste Biobrennstoffe*, Gülzow-Prüzen: Fachagentur Nachwachsende Rohstoffe e. V. (FNR), **2014**

[60] **EnergieAgentur.NRW**, *Holzpellets. Der Brennstoff der Zukunft.* Aktion Holzpellets, **2013**

[61] **Ericson C.**, *Real time collision detection*, San Francisco: Morgan Kaufmann Publishers, **2005**

[62] **Feng Y., Han K., Owen D.**, *A generic contact detection framework for cylindrical particles in discrete element modelling*, Computer Methods in Applied Mechanics and Engineering, vol. 315, pp. 632–651, DOI: 10.1016/j.cma.2016.11.001, **2017**

[63] **Ferrez J.-A.**, *Dynamic triangulations for efficient 3D simulation of granular materials*, PhD Thesis, Lausanne: University of Lausanne, **2001**

[64] **Ferziger J. H., Perić M., Street R. L.**, *Computational Methods for Fluid Dynamics*, Cham: Springer Nature Switzerland AG, DOI: 10.1007/978-3-319-99693-6, **2020**

[65] **Frye L., Peukert W.**, *Attrition of Bulk Solids in Pneumatic Conveying: Mechanisms and Material Properties*, Particulate Science and Technology, vol. 20, no. 4, pp. 267–282, DOI: 10.1080/02726350216187, **2002**

[66] **Frye L.**, "Chapter 28: A New Concept for Addressing Bulk Solids Attrition in Pneumatic Conveying", *Handbook of Powder Technology*, vol. 12, 07, pp. 1149–1218, DOI: 10.1016/S0167-3785(07)12031-5, **2007**

[67] **Gasterstadt J.**, *Die Experimentelle Untersuchung des Pneumatischen Fördervorganges*, Forschungsarbeiten Gebiete Ingenieurwesen, vol. 265, **1924**

[68] **Geldart D**, *Types of gas fluidization*, Powder Technology, vol. 7, no. 5, pp. 285–292, DOI: 10.1016/0032-5910(73)80037-3, **1973**

[69] **German Wood Fuel and Pellet Association**, *Recommendations for storage of wood pellets*, Berlin: German Wood Fuel and Pellet Association (DEPV), **2012**

[70] **Gidaspow D.**, *Multiphase flow and fluidization: Continuum and kinetic theory descriptions*, 2, London: Academic Press Limited, **1994**

[71] **Gilvari H., Jong W. D., Schott D. L.**, *Biomass Pellet Breakage : A Numerical Comparison Between Contact Models*, Proceedings of the 8th International Conference on Discrete Element Methods (DEM8), **2019**

[72] **Goldman R.**, *Understanding quaternions*, Graphical Models, vol. 73, no. 2, pp. 21–49, DOI: 10.1016/j.gmod.2010.10.004, **2011**

[73] **Graham S.**, *Degradation of biomass fuels during long term storage in indoor and outdoor environments*, PhD Thesis, Nottingham: University of Nottingham, **2015**

[74] **Graham S., Ogunfayo I., Hall M. R. et al.**, *Changes in mechanical properties of wood pellets during artificial degradation in a laboratory environment*, Fuel Processing Technology, vol. 148, pp. 395–402, DOI: 10.1016/j.fuproc.2016.03.020, **2016**

[75] **Grof Z., Kohout M., Štěpánek F.**, *Multi-scale simulation of needle-shaped particle breakage under uniaxial compaction*, Chemical Engineering Science, vol. 62, no. 5, pp. 1418–1429, DOI: 10.1016/j.ces.2006.11.033, **2007**

[76] **Grof Z., Schoellhammer C. M., Rajniak P., Štěpánek F.**, *Computational and experimental investigation of needle-shaped crystal breakage*, International Journal of Pharmaceutics, vol. 407, no. 1-2, pp. 12–20, DOI: 10.1016/j.ijpharm.2010.12.031, **2011**

[77] **Guo Y., Wassgren C., Hancock B. et al.**, *Predicting breakage of high aspect ratio particles in an agitated bed using the Discrete Element Method*, Chemical Engineering Science, vol. 158, no. April 2016, pp. 314–327, DOI: 10.1016/j.ces.2016.10.043, **2017**

[78] **Guo Y., Wassgren C., Ketterhagen W. et al.**, *A numerical study of granular shear flows of rod-like particles using the discrete element method*, Journal of Fluid Mechanics, vol. 713, pp. 1–26, DOI: 10.1017/jfm.2012.423, **2012**

[79] **Hadinoto K., Curtis J. S.**, *Reynolds number dependence of gas-phase turbulence in particle-laden flows: Effects of particle inertia and particle loading*, Powder Technology, vol. 195, no. 2, pp. 119–127, DOI: 10.1016/j.powtec.2009.05.022, **2009**

[80] **Hamilton W. R.**, *Elements of quaternions*, Camebridge, England: Cambridge University Press, **1866**

[81] **Han K., Feng Y., Owen D.**, *Performance comparisons of tree-based and cell-based contact detection algorithms*, Engineering Computations, vol. 24, no. 2, pp. 165–181, DOI: 10.1108/02644400710729554, **2007**

[82] **Harris C.**, *The application of size distribution equations to multi-event comminution process*, Transactions of the American Institute of Mining, Metallurgical and Petroleum Engineers, vol. 241, pp. 343–358, **1968**

[83] **Hartman M., Trnka O., Svoboda K.**, *Free Settling of Nonspherical Particles*, Industrial and Engineering Chemistry Research, vol. 33, no. 8, pp. 1979–1983, DOI: 10.1021/ie00032a012, **1994**

[84] **Henthorn K. H., Park K., Curtis J. S.**, *Measurement and Prediction of Pressure Drop in Pneumatic Conveying: Effect of Particle Characteristics, Mass Loading, and Reynolds Number*, Industrial & Engineering Chemistry Research, vol. 44, no. 14, pp. 5090–5098, DOI: 10.1021/ie049505e, **2005**

[85] **Hilbert J. D., Mills D.**, *Choices and decisions in pneumatic conveying system selection*, Chemistry Show Conference, **2005**

[86] **Hilgraf P.**, *Pneumatische Förderung*, Berlin, Heidelberg: Springer Berlin Heidelberg, DOI: 10.1007/978-3-662-58407-1, **2019**

[87] **Hilton J., Cleary P.**, *The influence of particle shape on flow modes in pneumatic conveying*, Chemical Engineering Science, vol. 66, no. 3, pp. 231–240, DOI: 10.1016/j.ces.2010.09.034, **2011**

[88] **Hilton J., Mason L., Cleary P.**, *Dynamics of gas-solid fluidised beds with non-spherical particle geometry*, Chemical Engineering Science, vol. 65, no. 5, pp. 1584–1596, DOI: 10.1016/j.ces.2009.10.028, **2010**

[89] **Hoerner S.**, *Fluid-dynamic drag*, New York: Hoerner Fluid Dynamics (published by the author), **1965**

[90] **Höhner D.**, *Experimentelle und numerische Untersuchungen zum Einfluss von Partikelgeometrie auf das mechanische Verhalten von Schüttgütern mit Hilfe der Diskreten Elemente Methode*, PhD Thesis, Bochum: Ruhr-University Bochum, **2014**

[91] **Höhner D., Wirtz S., Scherer V.**, *A numerical study on the influence of particle shape on hopper discharge within the polyhedral and multi-sphere discrete element method*, Powder Technology, vol. 226, pp. 16–28, DOI: 10.1016/j.powtec.2012.03.041, **2012**

[92] **Höhner D., Wirtz S., Scherer V.**, *Experimental and Numerical Investigation on the Discharge of Wood Pellets from Hopper with the Discrete Element Method*, ASME 2012 International Mechanical Engineering Congress & Exposition, Houston, pp. 1–11, **2012**

[93] **Höhner D., Wirtz S., Scherer V.**, *Experimental and numerical investigation on the influence of particle shape and shape approximation on hopper discharge using the discrete element method*, Powder Technology, vol. 235, pp. 614–627, DOI: 10.1016/j.powtec.2012.11.004, **2013**

[94] **Hold J.**, *Charakterisierung der Mischungs- und Segregations- vorgänge in Modellwirbelschichten mit mono- und bidispersen sphärischen Partikelsystemen : Ex-

perimente und gekoppelte CFD / DEM-Modellierung, PhD Thesis, Bochum: Ruhr-University Bochum, **2013**

[95] **Hölzer A., Sommerfeld M.**, *New simple correlation formula for the drag coefficient of non-spherical particles*, Powder Technology, vol. 184, no. 3, pp. 361–365, DOI: 10.1016/j.powtec.2007.08.021, **2008**

[96] **Hölzer A., Sommerfeld M.**, *Lattice Boltzmann simulations to determine drag, lift and torque acting on non-spherical particles*, Computers and Fluids, vol. 38, no. 3, pp. 572–589, DOI: 10.1016/j.compfluid.2008.06.001, **2009**

[97] **Hoomans B. P., Kuipers J. A., Briels W. J. et al.**, *Comments on the paper "Numerical simulation of the gas-solid flow in a fluidized bed by combining discrete particle method with computational fluid dynamics" by B. H. Xu and A. B. Yu*, Chemical Engineering Science, vol. 53, no. 14, pp. 2645–2646, DOI: 10.1016/S0009-2509(98)00085-2, **1998**

[98] **Hughes N. M., Shahi C., Pulkki R.**, *A Review of the Wood Pellet Value Chain, Modern Value/Supply Chain Management Approaches, and Value/Supply Chain Models*, Journal of Renewable Energy, vol. 2014, pp. 1–14, DOI: 10.1155/2014/654158, **2014**

[99] **IEA Bioenergy**, *Health and Safety Aspects of Solid Biomass Storage, Transportation and Feeding*, May, **2013**

[100] **Jones M. G.**, *The Influence of Bulk Particulate Properties on Pneumatic Conveying Performance*, PhD Thesis, London: Thames Polytechnic, **1988**

[101] **Jones M. G., Williams K. C.**, *Solids Friction Factors for Fluidized Dense-Phase Conveying*, Particulate Science and Technology, vol. 21, no. 1, pp. 45–56, DOI: 10.1080/02726350307495, **2003**

[102] **Kalman H., Hubert M., Grant E. et al.**, *Fatigue behavior of impact comminution and attrition units*, Powder Technology, vol. 146, no. 1-2, pp. 1–9, DOI: 10.1016/j.powtec.2004.07.011, **2004**

[103] **Kalman H., Rodnianski V., Haim M.**, *A new method to implement comminution functions into DEM simulation of a size reduction system due to particle-wall collisions*, Granular Matter, vol. 11, no. 4, pp. 253–266, DOI: 10.1007/s10035-009-0140-8, **2009**

[104] **Kalman H., Santo N., Tripathi N. M.**, *Particle velocity reduction in horizontal-horizontal bends of dilute phase pneumatic conveying*, Powder Technology, vol. 356, pp. 808–817, DOI: 10.1016/j.powtec.2019.08.081, **2019**

[105] **Kaltschmitt M., Hofbauer H., Hartmann H.**, *Energie aus Biomasse*, Berlin, Heidelberg: Springer Berlin Heidelberg, DOI: 10.1007/978-3-662-47438-9, **2016**

[106] **Kannan S., Gurusamy V., Nalini G.**, *A review on image segmentation techniques*, National Conference on Recent Trends and Research Issues in Computer Science, **2014**

[107] **Kennedy O.**, *Pneumatic conveying performance characteristics of bulk solids*, PhD Thesis, University of Wollongong, **1998**

[108] **Khan S., Ahmed V.**, *Smart Trends in Information Technology and Computer Communications, International Conference on Smart Trends for Information Technology and Computer Communications, Springer, Singapore*, vol. 628, August 2016, Singapore: Springer Singapore, DOI: 10.1007/978-981-10-3433-6, **2016**

[109] **Khanal M.**, *Simulation of Crushing Dynamics of an Aggregate-Matrix Composite by Compression and Impact Stressing*, PhD Thesis, Otto-von-Guericke-Universityof Magdeburg, **2005**

[110] **Khanal M., Schubert W., Tomas J.**, *Experiment and Simulation of Breakage of Particle Compounds under Compressive Loading*, Particulate Science and Technology, vol. 23, no. 4, pp. 387–394, DOI: 10.1080/02726350500213010, **2005**

[111] **Klinzing G. E., Rizk F., Marcus R., Leung L. S.**, *Pneumatic Conveying of Solids*, Springer Berlin Heidelberg, DOI: 10.1007/978-90-481-3609-4, **2010**

[112] **Klinzing G. E.**, *A review of pneumatic conveying status, advances and projections*, Powder Technology, vol. 333, pp. 78–90, DOI: 10.1016/j.powtec.2018.04.012, **2018**

[113] **Klinzing G. E.**, *Historical Review of Pneumatic Conveying*, KONA Powder and Particle Journal, vol. 35, no. 35, pp. 150–159, DOI: 10.14356/kona.2018010, **2018**

[114] **Kotzur B. A., Berry R. J., Bradley M., Dias G. C.**, *Influence of pellet length on breakage by impact*, 12th International Conference on Bulk Materials Storage, Handling and Transportation, Darwin, N.T., **2016**

[115] **Kotzur B. A., Bradley M. S., Berry R. J., Farnish R. J.**, *Breakage Characteristics of Granulated Food Products for Prediction of Attrition during Lean-Phase Pneumatic Conveying*, International Journal of Food Engineering, vol. 12, no. 9, pp. 835–850, DOI: 10.1515/ijfe-2016-0045, **2016**

[116] **Krüger B.**, *Experimentelle Charakterisierung und Modellierung der Flugbewegung von asphärischen Partikeln*, PhD Thesis, Bochum: Ruhr-University Bochum, **2015**

[117] **Kruggel-Emden H., Oschmann T.**, *Numerical study of rope formation and dispersion of non-spherical particles during pneumatic conveying in a pipe bend*, Powder Technology, vol. 268, pp. 219–236, DOI: 10.1016/j.powtec.2014.08.033, **2014**

[118] **Kruggel-Emden H., Simsek E., Rickelt S. et al.**, *Review and extension of normal force models for the Discrete Element Method*, Powder Technology, vol. 171, no. 3, pp. 157–173, DOI: 10.1016/j.powtec.2006.10.004, **2007**

[119] **Kruggel-Emden H., Sturm M., Wirtz S., Scherer V.**, *Selection of an appropriate time integration scheme for the discrete element method (DEM)*, Computers & Chemical Engineering, vol. 32, no. 10, pp. 2263–2279, DOI: 10.1016/j.compchemeng.2007.11.002, **2008**

[120] **Kruggel-Emden H.**, *Analysis and Improvement of the Time-Driven Discrete Element Method*, PhD Thesis, Bochum: Ruhr-University Bochum, **2007**

[121] **Krutikova O., Sisojevs A., Kovalovs M.**, *Creation of a Depth Map from Stereo Images of Faces for 3D Model Reconstruction*, Procedia Computer Science, vol. 104, no. December 2016, pp. 452–459, DOI: 10.1016/j.procs.2017.01.159, **2017**

[122] **Kuan B., Yang W., Schwarz M.**, *Dilute gas-solid two-phase flows in a curved 90-duct bend: CFD simulation with experimental validation*, Chemical Engineering Science, vol. 62, no. 7, pp. 2068–2088, DOI: 10.1016/j.ces.2006.12.054, **2007**

[123] **Kuang S., Li K., Zou R. et al.**, *Application of periodic boundary conditions to CFD-DEM simulation of gas-solid flow in pneumatic conveying*, Chemical Engineering Science, vol. 93, pp. 214–228, DOI: 10.1016/j.ces.2013.01.055, **2013**

[124] **Kuang S., Zhou M., Yu A.**, *CFD-DEM modelling and simulation of pneumatic conveying: A review*, Powder Technology, vol. 365, pp. 186–207, DOI: 10.1016/j.powtec.2019.02.011, **2020**

[125] **Kuipers J. B.**, *Quaternions and Rotation Sequences*, Princeton: Princeton University Press, DOI: 10.2307/3909157, **2000**

[126] **Kumar R., Ketterhagen W., Sarkar A. et al.**, *Breakage modeling of needle-shaped particles using the discrete element method*, Chemical Engineering Science: X, vol. 3, no. 100027, DOI: 10.1016/j.cesx.2019.100027, **2019**

[127] **Kuparinen K., Heinimö J., Vakkilainen E.**, *World's largest biofuel and pellet plants - geographic distribution, capacity share, and feedstock supply*, Biofuels, Bioproducts and Biorefining, vol. 8, no. 6, pp. 747–754, DOI: 10.1002/bbb.1516, **2014**

[128] **Kushimoto K., Ishihara S., Kano J.**, *Development of ADEM-CFD model for analyzing dynamic and breakage behavior of aggregates in wet ball milling*, Advanced Powder Technology, vol. 30, no. 6, pp. 1131–1140, DOI: 10.1016/j.apt.2019.03.008, **2019**

[129] **Kwade A., Schwedes J.**, "Chapter 6: Wet Grinding in Stirred Media Mills", *Handbook of Powder Technology*, vol. 12, 07, pp. 251–382, DOI: 10.1016/S0167-3785(07)12009-1, **2007**

[130] **Laín S., Sommerfeld M.**, *Numerical prediction of particle erosion of pipe bends*, Advanced Powder Technology, vol. 30, no. 2, pp. 366–383, DOI: 10.1016/j.apt.2018.11.014, **2019**

[131] **Laín S., Sommerfeld M.**, *Characterisation of pneumatic conveying systems using the Euler/Lagrange approach*, Powder Technology, vol. 235, pp. 764–782, DOI: 10.1016/j.powtec.2012.11.029, **2013**

[132] **Lee L. Y., Yong Quek T., Deng R. et al.**, *Pneumatic transport ofgranular materials through a 90-bend*, Chemical Engineering Science, vol. 59, no. 21, pp. 4637–4651, DOI: 10.1016/j.ces.2004.07.007, **2004**

[133] **Lerma-Arce V., Oliver-Villanueva J. V., Segura-Orenga G.**, *Influence of raw material composition of Mediterranean pinewood on pellet quality*, Biomass and Bioenergy, vol. 99, pp. 90–96, DOI: 10.1016/j.biombioe.2017.02.018, **2017**

[134] **Levy A.**, *Two-fluid approach for plug flow simulations in horizontal pneumatic conveying*, Powder Technology, vol. 112, no. 3, pp. 263–272, DOI: 10.1016/S0032-5910(00)00301-6, **2000**

[135] **Mahajan V. V., Nijssen T. M., Kuipers J., Padding J. T.**, *Non-spherical particles in a pseudo-2D fluidised bed: Modelling study*, Chemical Engineering Science, vol. 192, pp. 1105–1123, DOI: 10.1016/j.ces.2018.08.041, **2018**

[136] **Mahajan V. V., Padding J. T., Nijssen T. M. J. et al.**, *Nonspherical particles in a pseudo-2D fluidized bed: Experimental study*, AIChE Journal, vol. 64, no. 5, pp. 1573–1590, DOI: 10.1002/aic.16078, **2018**

[137] **Mandø M., Rosendahl L.**, *On the motion of non-spherical particles at high Reynolds number*, Powder Technology, vol. 202, no. 1-3, pp. 1–13, DOI: 10.1016/j.powtec.2010.05.001, **2010**

[138] **Manjula E., Ariyaratne W. H., Ratnayake C., Melaaen M. C.**, *A review of CFD modelling studies on pneumatic conveying and challenges in modelling*

offshore drill cuttings transport, Powder Technology, vol. 305, pp. 782–793, DOI: 10.1016/j.powtec.2016.10.026, **2017**

[139] **Masche M.**, *Milling and Physical Properties of Wood Pellets for Suspension-Fired Power Plants*, PhD Thesis, Technical University of Denmark, **2019**

[140] **Masche M., Puig-Arnavat M., Jensen P. A. et al.**, *From wood chips to pellets to milled pellets: The mechanical processing pathway of Austrian pine and European beech*, Powder Technology, vol. 350, pp. 134–145, DOI: 10.1016/j.powtec.2019.03.002, **2019**

[141] **Masche M., Puig-Arnavat M., Wadenbäck J. et al.**, *Wood pellet milling tests in a suspension-fired power plant*, Fuel Processing Technology, vol. 173, pp. 89–102, DOI: 10.1016/j.fuproc.2018.01.009, **2018**

[142] **Matsumoto M., Nishimura T.**, *Mersenne twister: a 623-dimensionally equidistributed uniform pseudo-random number generator*, ACM Transactions on Modeling and Computer Simulation, vol. 8, no. 1, pp. 3–30, DOI: 10.1145/272991.272995, **1998**

[143] **McGlinchey D., Cowell A., Knight E. A. et al.**, *Bend Pressure Drop Predictions Using the Euler-Euler Model in Dense Phase Pneumatic Conveying*, Particulate Science and Technology, vol. 25, no. 6, pp. 495–506, **2007**

[144] **McGlinchey D.**, *Bulk Solids Handling*, Oxford: Blackwell Publishing, **2008**

[145] **Mema I., Mahajan V. V., Fitzgerald B. W., Padding J. T.**, *Effect of lift force and hydrodynamic torque on fluidisation of non-spherical particles*, 12th International Conference on CFD in Oil & Gas. Metallurgical and Process Industries, DOI: 10.1016/j.ces.2018.10.009, **2017**

[146] **Meng Y., Zhang Z., Yin H., Ma T.**, *Automatic detection of particle size distribution by image analysis based on local adaptive canny edge detection and modified circular Hough transform*, Micron, vol. 106, no. August 2017, pp. 34–41, DOI: 10.1016/j.micron.2017.12.002, **2018**

[147] **Menter F.**, *Zonal Two Equation k-w Turbulence Models For Aerodynamic Flows*, 23rd Fluid Dynamics, Plasmadynamics, and Lasers Conference, Reston, Virigina, DOI: 10.2514/6.1993-2906, **1993**

[148] **Menter F.**, *Two-equation eddy-viscosity turbulence models for engineering applications*, AIAA Journal, vol. 32, no. 8, pp. 1598–1605, DOI: 10.2514/3.12149, **1994**

[149] **Meyer F., Vachier C.**, *On the Regularization of the Watershed Transform*, Advances in Imaging and Electron Physics, vol. 148, no. 07, pp. 193–249, DOI: 10.1016/S1076-5670(07)48003-X, **2007**

[150] **Mills D.**, *An investigation of the unstable region for dense phase conveying in sliding bed flow*, Granular Matter, vol. 6, no. 2-3, pp. 173–177, DOI: 10.1007/s10035-004-0162-1, **2004**

[151] **Mills D.**, *Pneumatic Conveying Design Guide*, Oxford: Elsevier, DOI: 10.1016/C2014-0-02678-0, **2016**

[152] **Mills D., Jones M. G., Agarwal V. K.**, *Handbook of Pneumatic Conveying Engineering*, New York: Marcel Dekker Inc., **2004**

[153] **Mina-Boac J., Maghirang R. G., Casada M. E.**, *Durability and Breakage of Feed Pellets during Repeated Elevator Handling*, ASABE Annual International Meeting, Oregon, **2006**

[154] **Miranda T., Montero I., Sepúlveda F. J. et al.**, *A review of pellets from different sources*, Materials, vol. 8, no. 4, pp. 1413–1427, DOI: 10.3390/ma8041413, **2015**

[155] **Muschelknauz E., Wojahn H.**, *Auslegung pneumatischer Förderanlagen*, Chemie Ingenieur Technik, vol. 46, no. 2, pp. 223–272, **1974**

[156] **Nan W., Wang Y., Wang J.**, *Numerical analysis on the fluidization dynamics of rodlike particles*, Advanced Powder Technology, vol. 27, no. 5, pp. 2265–2276, DOI: 10.1016/j.apt.2016.08.015, **2016**

[157] **Nezami E. G., Hashash Y. M., Zhao D., Ghaboussi J.**, *A fast contact detection algorithm for 3-D discrete element method*, Computers and Geotechnics, vol. 31, no. 7, pp. 575–587, DOI: 10.1016/j.compgeo.2004.08.002, **2004**

[158] **Norouzi H. R., Zarghami R., Sotudeh-Gharebagh R., Mostoufi N.**, *Coupled CFD-DEM Modeling - Formulation, Implementation and Application to Multiphase Flows*, Chichester, UK: John Wiley & Sons, Ltd., **2016**

[159] **Obernberger I., Thek G.**, *Physical characterisation and chemical composition of densified biomass fuels with regard to their combustion behaviour*, Biomass and Bioenergy, vol. 27, no. 6, pp. 653–669, DOI: 10.1016/j.biombioe.2003.07.006, **2004**

[160] **Obernberger I., Thek G.**, *The Pellet Handbook - the production and thermal utilisation of biomass pellets*, London: Earthscan Ltd., **2010**

[161] **Ochkin-König O., Dosta M., Heinrich S.**, *Investigation of breakage and attrition behaviour of wood pellets*, International Conference on Conveying and Handling of Particulate Solids, London, **2018**

[162] **Oertel H.**, *Strömungsmechanik - Methoden und Phänomene*, Springer Verlag Berlin Heidelberg, **1995**

[163] **Ogarko V., Luding S.**, *Data structures and algorithms for contact detection in numerical simulation of discrete particle systems*, World Congress Particle Technology 6, Nuremberg, **2010**

[164] **Orozovic O., Lavrinec A., Alkassar Y. et al.**, *On the kinematics of horizontal slug flow pneumatic conveying and the relationship between slug length, porosity, velocities and stationary layers*, Powder Technology, vol. 351, pp. 84–91, DOI: 10.1016/j.powtec.2019.04.017, **2019**

[165] **Oschmann T., Hold J., Kruggel-Emden H.**, *Numerical investigation of mixing and orientation of non-spherical particles in a model type fluidized bed*, Powder Technology, vol. 258, pp. 304–323, DOI: 10.1016/j.powtec.2014.03.046, **2014**

[166] **Oschmann T., Vollmari K., Kruggel-Emden H., Wirtz S.**, *Numerical Investigation of the Mixing of Non-spherical Particles in Fluidized Beds and during Pneumatic Conveying*, Procedia Engineering, vol. 102, pp. 976–985, DOI: 10.1016/j.proeng.2015.01.220, **2015**

[167] **Oveisi E., Lau A., Sokhansanj S. et al.**, *Breakage behavior of wood pellets due to free fall*, Powder Technology, vol. 235, pp. 493–499, DOI: 10.1016/j.powtec.2012.10.022, **2013**

[168] **Pahk J. B., Klinzing G. E.**, *Comparison of flow characteristics for dilute phase pneumatic conveying for two different plastic pellets*, Journal of the Chinese Institute of Chemical Engineers, vol. 39, no. 2, pp. 143–150, DOI: 10.1016/j.jcice.2007.12.005, **2008**

[169] **Patro P., Dash S. K.**, *Two-fluid modeling of turbulent particle-gas suspensions in vertical pipes*, Powder Technology, vol. 264, pp. 320–331, DOI: 10.1016/j.powtec.2014.05.048, **2014**

[170] **Pocock J., Veasey T., Tavares L., King R.**, *The effect of heating and quenching on grinding characteristics of quartzite*, Powder Technology, vol. 95, no. 2, pp. 137–142, DOI: 10.1016/S0032-5910(97)03333-0, **1998**

[171] **Portnikov D., Peisakhov R., Gabrieli G. O., Kalman H.**, *Selection function of particles under impact loads: The effect of collision angle*, Particulate Science

and Technology, vol. 36, no. 4, pp. 420–426, DOI: 10.1080/02726351.2017.1346022, **2018**

[172] **Portnikov D., Santo N., Kalman H.**, *Simplified model for particle collision related to attrition in pneumatic conveying*, Advanced Powder Technology, vol. 31, no. 1, pp. 359–369, DOI: 10.1016/j.apt.2019.10.028, **2020**

[173] **Proskurina S., Junginger M., Heinimö J. et al.**, *Global biomass trade for energy- Part 2: Production and trade streams of wood pellets, liquid biofuels, charcoal, industrial roundwood and emerging energy biomass*, Biofuels, Bioproducts and Biorefining, pp. 1–17, DOI: 10.1002/bbb.1858, **2018**

[174] **Proskurina S., Junginger M., Heinimö J., Vakkilainen E.**, *Global biomass trade for energy - Part 1: Statistical and methodological considerations*, Biofuels, Bioproducts and Biorefining, DOI: 10.1002/bbb.1841, **2017**

[175] **Pu W., Zhao C., Xiong Y. et al.**, *Numerical simulation on dense phase pneumatic conveying of pulverized coal in horizontal pipe at high pressure*, Chemical Engineering Science, vol. 65, no. 8, pp. 2500–2512, DOI: 10.1016/j.ces.2009.12.025, **2010**

[176] **Rao A., Curtis J. S., Hancock B. C., Wassgren C.**, *Numerical simulation of dilute turbulent gas-particle flow with turbulence modulation*, AIChE Journal, vol. 58, no. 5, pp. 1381–1396, DOI: 10.1002/aic.12673, **2012**

[177] **Richter A., Nikrityuk P. A.**, *Drag forces and heat transfer coefficients for spherical, cuboidal and ellipsoidal particles in cross flow at sub-critical Reynolds numbers*, International Journal of Heat and Mass Transfer, vol. 55, no. 4, pp. 1343–1354, DOI: 10.1016/j.ijheatmasstransfer.2011.09.005, **2012**

[178] **Rodnianski V., Levy A., Kalman H.**, *A new method for simulation of comminution process in jet mills*, Powder Technology, vol. 343, pp. 867–879, DOI: 10.1016/j.powtec.2018.11.021, **2019**

[179] **Rodríguez-Quiñonez J., Sergiyenko O., Flores-Fuentes W. et al.**, *Improve a 3D distance measurement accuracy in stereo vision systems using optimization methods' approach*, Opto-Electronics Review, vol. 25, no. 1, pp. 24–32, DOI: 10.1016/j.opelre.2017.03.001, **2017**

[180] **Roessler T., Katterfeld A.**, *Scaling of the angle of repose test and its influence on the calibration of DEM parameters using upscaled particles*, Powder Technology, vol. 330, pp. 58–66, DOI: 10.1016/j.powtec.2018.01.044, **2018**

[181] **Roessler T., Katterfeld A.**, *DEM parameter calibration of cohesive bulk materials using a simple angle of repose test*, Particuology, vol. 45, pp. 105–115, DOI: 10.1016/j.partic.2018.08.005, **2019**

[182] **Roessler T., Richter C., Katterfeld A., Will F.**, *Development of a standard calibration procedure for the DEM parameters of cohesionless bulk materials - part I: Solving the problem of ambiguous parameter combinations*, Powder Technology, vol. 343, pp. 803–812, DOI: 10.1016/j.powtec.2018.11.034, **2019**

[183] **Rozenblat Y., Grant E., Levy A. et al.**, *Selection and breakage functions of particles under impact loads*, Chemical Engineering Science, vol. 71, pp. 56–66, DOI: 10.1016/j.ces.2011.12.012, **2012**

[184] **Rozenblat Y., Levy A., Kalman H., Tomas J.**, *Impact velocity and compression force relationship - Equivalence function*, Powder Technology, vol. 235, pp. 756–763, DOI: 10.1016/j.powtec.2012.11.011, **2013**

[185] **Rozenblat Y., Levy A., Kalman H. et al.**, *A model for particle fatigue due to impact loads*, Powder Technology, vol. 239, pp. 199–207, DOI: 10.1016/j.powtec.2013.01.059, **2013**

[186] **Rozenblat Y., Portnikov D., Levy A. et al.**, *Strength distribution of particles under compression*, Powder Technology, vol. 208, no. 1, pp. 215–224, DOI: 10.1016/j.powtec.2010.12.023, **2011**

[187] **Salman A., Biggs C., Fu J. et al.**, *An experimental investigation of particle fragmentation using single particle impact studies*, Powder Technology, vol. 128, no. 1, pp. 36–46, DOI: 10.1016/S0032-5910(02)00151-1, **2002**

[188] **Salman A., Hounslow M., Verba A.**, *Particle fragmentation in dilute phase pneumatic conveying*, Powder Technology, vol. 126, no. 2, pp. 109–115, DOI: 10.1016/S0032-5910(02)00048-7, **2002**

[189] **Sanchez L., Vasquez N., Klinzing G. E., Dhodapkar S.**, *Characterization of bulk solids to assess dense phase pneumatic conveying*, Powder Technology, vol. 138, no. 2-3, pp. 93–117, DOI: 10.1016/j.powtec.2003.08.061, **2003**

[190] **Sanjeevi S. K., Padding J. T.**, *On the orientational dependence of drag experienced by spheroids*, Journal of Fluid Mechanics, vol. 820, DOI: 10.1017/jfm.2017.239, **2017**

[191] **Santo N., Portnikov D., Eshel I. et al.**, *Experimental study on particle steady state velocity distribution in horizontal dilute phase pneumatic conveying*, Chemical Engineering Science, vol. 187, pp. 354–366, DOI: 10.1016/j.ces.2018.04.058, **2018**

[192] **Santo N., Portnikov D., Tripathi N. M., Kalman H.**, *Experimental study on the particle velocity development profile and acceleration length in horizontal dilute phase pneumatic conveying systems*, Powder Technology, vol. 339, pp. 368–376, DOI: 10.1016/j.powtec.2018.07.074, **2018**

[193] **Santos S. M., Tambourgi E. B., Fernandes F. A. N. et al.**, *Dilute-phase pneumatic conveying of polystyrene particles: pressure drop curve and particle distribution over the pipe cross-section*, Brazilian Journal of Chemical Engineering, vol. 28, no. 1, pp. 81–88, DOI: 10.1590/S0104-66322011000100010, **2011**

[194] **Schäfer J., Dippel S., Wolf D. E.**, *Force Schemes in Simulations of Granular Materials*, Journal de Physique I, vol. 6, no. 1, pp. 5–20, DOI: 10.1051/jp1:1996129, **1996**

[195] **Schallert R., Levy E.**, *Effect of a combination of two elbows on particle roping in pneumatic conveying*, Powder Technology, vol. 107, no. 3, pp. 226–233, DOI: 10.1016/S0032-5910(99)00189-8, **2000**

[196] **Schiller L., Naumann A.**, *A drag coefficient correlation*, Journal of the Association of German Engineers (VDI), vol. 77, pp. 318–320, **1935**

[197] **Schulz D., Schwindt N., Schmidt E. et al.**, *Investigation of the dust release from bulk material undergoing various mechanical processes using a coupled DEM/CFD approach*, Powder Technology, vol. 355, pp. 37–56, DOI: 10.1016/j.powtec.2019.07.005, **2019**

[198] **Schumann Jr. R.**, *Principles of comminution, I. Size distribution and surface calculations*, AIME Technical Paper, vol. 1189, pp. 1–11, **1940**

[199] **Schwarze R.**, *CFD-Modellierung*, Berlin, Heidelberg: Springer Berlin Heidelberg, DOI: 10.1007/978-3-642-24378-3, **2013**

[200] **Sharma K., Mallick S., Mittal A., Pan R.**, *On developing improved modelling for particle velocity and solids friction for fluidized dense-phase pneumatic transport systems*, Powder Technology, vol. 332, pp. 41–55, DOI: 10.1016/j.powtec.2018.03.039, **2018**

[201] **Shimizu Y.**, *Three-dimensional simulation using fixed coarse-grid thermal-fluid scheme and conduction heat transfer scheme in distinct element method*, Powder Technology, vol. 165, no. 3, pp. 140–152, DOI: 10.1016/j.powtec.2006.04.003, **2006**

[202] **Shukla P. R., Skea J., Calvo Buendia E. et al.**, *Climate Change and Land: an IPCC special report on climate change, desertification, land degradation, sus-*

tainable land management, food security, and greenhouse gas fluxes in terrestrial ecosystems, Intergovernmental Panel on Climate Change (IPPC), **2019**

[203] **Simsek E.**, *Mischung und Segregation auf Rostsystemen : Experimentelle Untersuchung und numerische Simulation mit Hilfe der Diskrete Elemente Methode*, PhD Thesis, Bochum: Ruhr-University Bochum, **2010**

[204] **Solnordal C. B., Wong C. Y., Boulanger J.**, *An experimental and numerical analysis of erosion caused by sand pneumatically conveyed through a standard pipe elbow*, Wear, vol. 336-337, pp. 43–57, DOI: 10.1016/j.wear.2015.04.017, **2015**

[205] **Sørensen H., Rosendahl L., Yin C., Mandø M.**, *Settling of a cylindrical particle in a stagnant fluid*, 6th International Conference on Multiphase Flow, pp. 1–9, July, **2007**

[206] **Spettl A., Dosta M., Antonyuk S. et al.**, *Statistical investigation of agglomerate breakage based on combined stochastic microstructure modeling and DEM simulations*, Advanced Powder Technology, vol. 26, no. 3, pp. 1021–1030, DOI: 10.1016/j.apt.2015.04.011, **2015**

[207] **Stamboliadis E. T.**, *The energy distribution theory of comminution specific surface energy, mill efficiency and distribution mode*, Minerals Engineering, vol. 20, no. 2, pp. 140–145, DOI: 10.1016/j.mineng.2006.07.009, **2007**

[208] **Stieß M.**, *Mechanische Verfahrenstechnik - Partikeltechnologie 1*, Berlin, Heidelberg: Springer Berlin Heidelberg, DOI: 10.1007/978/3-540-32552-9, **2009**

[209] **Takeuchi H., Nakamura H., Watano S.**, *Numerical simulation of particle breakage in dry impact pulverizer*, AIChE Journal, vol. 59, no. 10, pp. 3601–3611, DOI: 10.1002/aic.14096, **2013**

[210] **Tang C., Liu H.**, "Chapter 16: Numerical Investigation of Particle Breakage as Applied to Mechanical Crushing", *Handbook of Powder Technology*, vol. 12, 07, pp. 661–739, DOI: 10.1016/S0167-3785(07)12019-4, **2007**

[211] **Tavares L.**, *Optimum routes for particle breakage by impact*, Powder Technology, vol. 142, no. 2-3, pp. 81–91, DOI: 10.1016/j.powtec.2004.03.014, **2004**

[212] **Theidel M.**, *DEM-CFD Simulation dichter Fluid-Feststoff-Strömungen: pneumatischer Transport von Schüttgut*, PhD Thesis, Bochum: Ruhr-University Bochum, **2011**

[213] **Timmerman P., Weele J. P. van der**, *On the rise and fall of a ball with linear or quadratic drag*, American Journal of Physics, vol. 67, no. 6, pp. 538–546, DOI: 10.1119/1.19320, **1999**

[214] **Tripathi N. M., Levy A., Kalman H.**, *Acceleration pressure drop analysis in horizontal dilute phase pneumatic conveying system*, Powder Technology, vol. 327, pp. 43–56, DOI: 10.1016/j.powtec.2017.12.045, **2018**

[215] **Tripathi N. M., Portnikov D., Levy A., Kalman H.**, *Bend pressure drop in horizontal and vertical dilute phase pneumatic conveying systems*, Chemical Engineering Science, vol. 209, p. 115228, DOI: 10.1016/j.ces.2019.115228, **2019**

[216] **Tripathi N. M., Santo N., Kalman H., Levy A.**, *Experimental analysis of particle velocity and acceleration in vertical dilute phase pneumatic conveying*, Powder Technology, vol. 330, no. 2017, pp. 239–251, DOI: 10.1016/j.powtec.2018.02.017, **2018**

[217] **Tripathi N. M., Santo N., Levy A., Kalman H.**, *Experimental analysis of velocity reduction in bends related to vertical pipes in dilute phase pneumatic conveying*, Powder Technology, vol. 345, pp. 190–202, DOI: 10.1016/j.powtec.2019.01.001, **2019**

[218] **Tripathi N., Sharma A., Mallick S. S., Wypych P. W.**, *Energy loss at bends in the pneumatic conveying of fly ash*, Particuology, vol. 21, pp. 65–73, DOI: 10.1016/j.partic.2014.09.003, **2015**

[219] **Tsuji Y., Kawaguchi T., Tanaka T.**, *Discrete particle simulation of two-dimensional fluidized bed*, Powder Technology, vol. 77, no. 1, pp. 79–87, DOI: 10.1016/0032-5910(93)85010-7, **1993**

[220] **Tsuji Y., Tanaka T., Ishida T.**, *Lagrangian numerical simulation of plug flow of cohesionless particles in a horizontal pipe*, Powder Technology, vol. 71, no. 3, pp. 239–250, DOI: 10.1016/0032-5910(92)88030-L, **1992**

[221] **Uasuf A.**, *Economic and environmental assessment of an international wood pellets supply chain: a case study of wood pellets export from northeast Argentina to Europe*, PhD Thesis, Albert Ludwig University of Freiburg, **2010**

[222] **Uzi A., Ben Ami Y., Levy A.**, *Erosion prediction of industrial conveying pipelines*, Powder Technology, vol. 309, pp. 49–60, DOI: 10.1016/j.powtec.2016.12.087, **2017**

[223] **Uzi A., Kalman H., Levy A.**, *A novel particle attrition model for conveying systems*, Powder Technology, vol. 298, pp. 30–41, DOI: 10.1016/j.powtec.2016.05.014, **2016**

[224] **VDI Gesellschaft**, *VDI-Wärmeatlas*, *VDI Gesellschaft Verfahrenstechnik und Chemieingenieurwesen*, vol. 11, Berlin: Springer Berlin Heidelberg, DOI: 10.1007/978-3-642-19981-3, **2013**

[225] **Vásquez N., Jacob K., Cocco R. et al.**, *Visual analysis of particle bouncing and its effect on pressure drop in dilute phase pneumatic conveying*, Powder Technology, vol. 179, no. 3, pp. 170–175, DOI: 10.1016/j.powtec.2007.06.015, **2008**

[226] **Vogel L., Peukert W.**, *Characterisation of Grinding-Relevant Particle Properties by Inverting a Population Balance Model*, Particle & Particle Systems Characterization, vol. 19, no. 3, p. 149, DOI: 10.1002/1521-4117(200207)19:3<149::AID-PPSC149>3.0.CO;2-8, **2002**

[227] **Vollmari K., Kruggel-Emden H.**, *Numerical and experimental analysis of particle residence times in a continuously operated dual-chamber fluidized bed*, Powder Technology, vol. 338, pp. 625–637, DOI: 10.1016/j.powtec.2018.07.061, **2018**

[228] **Wang Y., Williams K., Jones M., Chen B.**, *CFD simulation methodology for gas-solid flow in bypass pneumatic conveying - A review*, Applied Thermal Engineering, vol. 125, pp. 185–208, DOI: 10.1016/j.applthermaleng.2017.05.063, **2017**

[229] **Weber M.**, *Strömungs-Fördertechnik*, Mainz: Krausskopf Verlag, **1974**

[230] **Weng J., Cohen P., Herniou M.**, *Camera calibration with distortion models and accuracy evaluation*, IEEE Transactions on Pattern Analysis and Machine Intelligence, vol. 14, no. 10, pp. 965–980, DOI: 10.1109/34.159901, **1992**

[231] **Whitmore S. A.**, *Closed-form integrator for the quaternion (euler angle) kinematics equations*, *New York*, Washington, D.C.: The United States of America as represented by the Administrator of the National Aeronautics and Space Administration, **1998**

[232] **Wiese J., Wirtz S., Scherer V., Behr H. M.**, *Einfluss der Lagerraumgeometrie und des Befüllvorgangs auf Feinanteile und Längenverteilung von Holzpellets*, Fachmagazin Pellets, no. 02-15, pp. 30–33, **2015**

[233] **Wiese J.**, *DEM/CFD-Simulation und experimentelle Untersuchungen von Holzpelletfeuerungen*, PhD Thesis, Bochum: Ruhr-University Bochum, **2015**

[234] **Wilcox D. C.**, *Formulation of the k-w Turbulence Model Revisited*, AIAA Journal, vol. 46, no. 11, pp. 2823–2838, DOI: 10.2514/1.36541, **2008**

[235] **Williams J. R., Perkins E., Cook B.**, *A contact algorithm for partitioning N arbitrary sized objects*, Engineering Computations, vol. 21, pp. 235–248, DOI: 10.1108/02644400410519767, **2004**

[236] **Williams K. C.**, *Dense Phase Pneumatic Conveying of Powders: Design Aspects and Phenomena*, PhD Thesis, Newcastle: University of Newcastle, **2008**

[237] **Williams S. R., Philipse A. P.**, *Random packings of spheres and spherocylinders simulated by mechanical contraction*, Physical Review E - Statistical Physics, Plasmas, Fluids, and Related Interdisciplinary Topics, vol. 67, no. 5, p. 9, DOI: 10.1103/PhysRevE.67.051301, **2003**

[238] **Wissing F., Wirtz S., Scherer V.**, *Simulating municipal solid waste incineration with a DEM/CFD method - Influences of waste properties, grate and furnace design*, Fuel, vol. 206, pp. 638–656, DOI: 10.1016/j.fuel.2017.06.037, **2017**

[239] **Wissing F.**, *DEM/CFD-Simulation und experimentelle Untersuchung der Transportvorgänge in Abfallverbrennungsanlagen*, PhD Thesis, Bochum: Ruhr-University Bochum, **2018**

[240] **Wypych P. W., Yi J.**, *Minimum transport boundary for horizontal dense-phase pneumatic conveying of granular materials*, Powder Technology, vol. 129, no. 1-3, pp. 111–121, DOI: 10.1016/S0032-5910(02)00224-3, **2003**

[241] **Xu B., Yu A.**, *Numerical simulation of the gas-solid flow in a fluidized bed by combining discrete particle method with computational fluid dynamics*, Chemical Engineering Science, vol. 52, no. 16, pp. 2785–2809, DOI: 10.1016/S0009-2509(97)00081-X, **1997**

[242] **Xu L., Zhang Q., Zheng J., Zhao Y.**, *Numerical prediction of erosion in elbow based on CFD-DEM simulation*, Powder Technology, vol. 302, pp. 236–246, DOI: 10.1016/j.powtec.2016.08.050, **2016**

[243] **Yang D., Xing B., Li J. et al.**, *Experiment and simulation analysis of the suspension behavior of large (5-30 mm) nonspherical particles in vertical pneumatic conveying*, Powder Technology, vol. 354, pp. 442–455, DOI: 10.1016/j.powtec.2019.06.023, **2019**

[244] **Yao Y., Wu W., Yang T. et al.**, *Head rice rate measurement based on concave point matching*, Scientific Reports, vol. 7, no. 1, p. 41353, DOI: 10.1038/srep41353, **2017**

[245] **Yilmaz A., Levy E. K.**, *Formation and dispersion of ropes in pneumatic conveying*, Powder Technology, vol. 114, no. 1-3, pp. 168–185, DOI: 10.1016/S0032-5910(00)00319-3, **2001**

[246] **Yin C., Rosendahl L., Kær S. K., Sørensen H.**, *Corrigendum to 'Modelling the motion of cylindrical particles in a nonuniform flow' [Chem. Eng. Sci. 58 (2003) 3489-3498]*, Chemical Engineering Science, vol. 66, no. 1, p. 117, DOI: 10.1016/j.ces.2010.10.021, **2011**

[247] **Yin C., Rosendahl L., Knudsen Kær S., Sørensen H.**, *Modelling the motion of cylindrical particles in a nonuniform flow*, Chemical Engineering Science, vol. 58, no. 15, pp. 3489–3498, DOI: 10.1016/S0009-2509(03)00214-8, **2003**

[248] **Zafari S., Eerola T., Sampo J. et al.**, "Segmentation of Partially Overlapping Nanoparticles Using Concave Points", *Advances in Visual Computing. ISVC 2015. Lecture Notes in Computer Science*, vol. 9474, pp. 187–197, DOI: 10.1007/978-3-319-27857-5_17, **2015**

[249] **Zastawny M., Mallouppas G., Zhao F., Wachem B. van**, *Derivation of drag and lift force and torque coefficients for non-spherical particles in flows*, International Journal of Multiphase Flow, vol. 39, pp. 227–239, DOI: 10.1016/j.ijmultiphaseflow.2011.09.004, **2012**

[250] **Zhang Y.**, *Mechanical behavior of granular material considering particle breakage*, PhD Thesis, Metz: Université de Lorraine, **2018**

[251] **Zhao J., Shan T.**, *Coupled CFD-DEM simulation of fluid-particle interaction in geomechanics*, Powder Technology, vol. 239, pp. 248–258, DOI: 10.1016/j.powtec.2013.02.003, **2013**

[252] **Zheng J., Hryciw R. D.**, *Segmentation of contacting soil particles in images by modified watershed analysis*, Computers and Geotechnics, vol. 73, pp. 142–152, DOI: 10.1016/j.compgeo.2015.11.025, **2016**

[253] **Zhong W., Yu A., Liu X. et al.**, *DEM/CFD-DEM Modelling of Non-spherical Particulate Systems: Theoretical Developments and Applications*, Powder Technology, vol. 302, pp. 108–152, DOI: 10.1016/j.powtec.2016.07.010, **2016**

[254] **Zhou J.-W., Han X., Jing S., Liu Y.**, *Efficiency and stability of lump coal particles swirling flow pneumatic conveying system*, Chemical Engineering Research and Design, vol. 157, pp. 92–103, DOI: 10.1016/j.cherd.2020.03.006, **2020**

[255] **Zhou J.-W., Liu Y., Du C. et al.**, *Numerical study of coarse coal particle breakage in pneumatic conveying*, Particuology, vol. 38, pp. 204–214, DOI: 10.1016/j.partic.2017.07.003, **2018**

[256] **Zhou J.-W., Liu Y., Liu S.-Y. et al.**, *Effects of particle shape and swirling intensity on elbow erosion in dilute-phase pneumatic conveying*, Wear, vol. 380-381, pp. 66–77, DOI: 10.1016/j.wear.2017.03.009, **2017**

[257] **Zhou J.-W., Xu L.-g., Du C.-l.**, *Prediction of lump coal particle pickup velocity in pneumatic conveying*, Powder Technology, vol. 343, pp. 599–606, DOI: 10.1016/j.powtec.2018.11.090, **2019**

[258] **Zhou Y. C., Xu B. H., Yu A. B., Zulli P.**, *Numerical investigation of the angle of repose of monosized spheres*, Physical Review E, vol. 64, no. 2, p. 021301, DOI: 10.1103/PhysRevE.64.021301, **2001**

[259] **Zhou Y., Wright B., Yang R. et al.**, *Rolling friction in the dynamic simulation of sandpile formation*, Physica A: Statistical Mechanics and its Applications, vol. 269, no. 2, pp. 536–553, DOI: 10.1016/S0378-4371(99)00183-1, **1999**

[260] **Zhou Y., Xu B., Yu A., Zulli P.**, *An experimental and numerical study of the angle of repose of coarse spheres*, Powder Technology, vol. 125, no. 1, pp. 45–54, DOI: 10.1016/S0032-5910(01)00520-4, **2002**

[261] **Zhou Z. Y., Liu S., Zou R. P. et al.**, *Numerical investigation of piling and hopper flow ellipsoidal particles*, Chemeca 2011: Engineering a Better World, Sydney Hilton Hotel, NSW, Australia, pp. 2890–2899, **2011**

[262] **Zhu H., Zhou Z., Yang R., Yu A.**, *Discrete particle simulation of particulate systems: Theoretical developments*, Chemical Engineering Science, vol. 62, no. 13, pp. 3378–3396, DOI: 10.1016/j.ces.2006.12.089, **2007**

[263] **Zhu H., Zhou Z., Yang R., Yu A.**, *Discrete particle simulation of particulate systems: A review of major applications and findings*, Chemical Engineering Science, vol. 63, no. 23, pp. 5728–5770, DOI: 10.1016/j.ces.2008.08.006, **2008**

[264] **Zhu K., Wong C. K., Madhusudana Rao S., Wang C.-H.**, *Pneumatic conveying of granular solids in horizontal and inclined pipes*, AIChE Journal, vol. 50, no. 8, pp. 1729–1745, DOI: 10.1002/aic.10172, **2004**

Own publications

[a] **Jägers J., Spatz P., Wirtz S., Scherer V.**, *Analysis of wood pellet degradation characteristics based on single particle impact tests*, Powder Technology, vol. 378, pp. 704–715, DOI: 10.1016/j.powtec.2020.10.017, **2021**

[b] **Jägers J., Wirtz S., Scherer V.**, *Experimentelle und numerische Analyse des Abrieb- und Bruchverhaltens von Holzpellets während pneumatischer Transportvorgänge*, Jahrestreffen der ProcessNet-Fachgruppen "Agglomeration und Schüttguttechnik" und "Zerkleinern und Klassieren", Neuss, March, **2018**

[c] **Jägers J., Wirtz S., Scherer V.**, *Analysis of wood pellet breakage during pneumatic delivery processes*, European Pellet Conference 2019, Wels, February, **2019**

[d] **Jägers J., Wirtz S., Scherer V.**, *Erfahrung hat den Bogen raus! Wissenschaftliche Untersuchung zum Transport von Holzpellets in das Lager*, pelletmagazin II/2019, pp. 26–28, Berlin, **2019**

[e] **Jägers J., Wirtz S., Scherer V.**, *Investigation of wood pellet breakage during pneumatic delivery processes*, 13th International Conference on Bulk Materials Storage, Handling & Transportation, Surfers Paradise, QLD, Australia, June, **2019**

[f] **Jägers J., Wirtz S., Scherer V.**, *Untersuchung der Zerkleinerungswirkung pneumatischer Transportvorgänge auf Holzpellets*, 29. Deutscher Flammentag, Bochum, September, **2019**

[g] **Jägers J., Wirtz S., Scherer V.**, *An automated and continuous method for the optical measurement of wood pellet size distribution and the gravimetric determination of fines*, Powder Technology, vol. 367, pp. 681–688, DOI: 10.1016/j.powtec.2020.04.023, **2020**

[h] **Jägers J., Wirtz S., Scherer V.**, *Erfahrung hat den Bogen raus!*, Brennstoffspiegel & Mineralölrundschau, no. 03, pp. 47–49, **2020**

[i] **Jägers J., Wirtz S., Scherer V.**, *Transport mit Druckluft - Einfluss von Betriebsbedingungen auf die Qualität von Holzpellets*, Moderne Gebäudetechnik - Das Objektgeschäft 2020, pp. 28–30, Berlin, **2020**

[j] **Jägers J., Wirtz S., Scherer V., Behr H. M.**, *Experimental analysis of wood pellet degradation during pneumatic conveying processes*, Powder Technology, vol. 359, pp. 282–291, DOI: 10.1016/j.powtec.2019.10.004, **2020**

Curriculum Vitae

Julian Jägers
Geboren am 9. November 1991 in Dinslaken
Verheiratet

Berufserfahrung

Seit 06/2016	Wissenschaftlicher Mitarbeiter an der Ruhr-Universität Bochum Lehrstuhl für Energieanlagen und Energieprozesstechnik
10/2015 - 03/2016	Studentische Hilfskraft an der Ruhr-Universität Bochum Lehrstuhl für Energieanlagen und Energieprozesstechnik
02/2013 - 07/2015	Werkstudent am Gas- und Wärme-Institut Essen e.V. Abteilung Industrie- und Feuerungstechnik
07/2011 - 09/2011	Studienbegleitendes Praktikum Borsig Service GmbH, Gladbeck

Hochschulausbildung

04/2015 - 03/2016	Studium des Maschinenbaus an der Ruhr-Universität Bochum Vertiefungsschwerpunkt: Energie- und Verfahrenstechnik Abschluss: Master of Science (M.Sc.)
10/2013 - 02/2014	Erasmus-Auslandssemester an der University of Sheffield (GB) Department of Chemical and Process Engineering
10/2011 - 03/2015	Studium des Maschinenbaus an der Ruhr-Universität Bochum Vertiefungsschwerpunkt: Energie- und Verfahrenstechnik Abschluss: Bachelor of Science (B.Sc.)

Schulausbildung

08/2002 - 07/2011	Städtisches Ratsgymnasium, Gladbeck Abschluss: Allgemeine Hochschulreife
08/1998 - 07/2002	Josefschule - kath. Grundschule, Gladbeck

www.ingramcontent.com/pod-product-compliance
Ingram Content Group UK Ltd.
Pitfield, Milton Keynes, MK11 3LW, UK
UKHW022001190726
13853UKWH00004B/1661